Game Theory: A Very Short Introduction

Very Short Introductions available now:

Available soon:

EXPRESSIONISM
  Katerina Reed-Tsocha
GALAXIES John Gribbin
GEOGRAPHY John Matthews and
  David Herbert
GERMAN LITERATURE
  Nicholas Boyle
HIV/AIDS Alan Whiteside
THE MEANING OF LIFE
  Terry Eagleton

MEMORY Jonathan Foster
MODERN CHINA
  Rana Mitter
NUCLEAR WEAPONS
  Joseph M. Siracusa
QUAKERISM Pink Dandelion
SCIENCE AND
  RELIGION Thomas Dixon
SEXUALITY
  Véronique Mottier

For more information visit our web site
www.oup.co.uk/general/vsi/

Ken Binmore

GAME THEORY

A Very Short Introduction

OXFORD
UNIVERSITY PRESS

OXFORD
UNIVERSITY PRESS

Great Clarendon Street, Oxford ox2 6dp

Oxford University Press is a department of the University of Oxford.
It furthers the University's objective of excellence in research, scholarship,
and education by publishing worldwide in

Oxford New York

Auckland Cape Town Dar es Salaam Hong Kong Karachi
Kuala Lumpur Madrid Melbourne Mexico City Nairobi
New Delhi Shanghai Taipei Toronto

With offices in

Argentina Austria Brazil Chile Czech Republic France Greece
Guatemala Hungary Italy Japan Poland Portugal Singapore
South Korea Switzerland Thailand Turkey Ukraine Vietnam

Oxford is a registered trade mark of Oxford University Press
in the UK and in certain other countries

Published in the United States
by Oxford University Press Inc., New York

British Library Cataloguing in Publication Data

Data available

Library of Congress Cataloging in Publication Data

Data available

ISBN 978-0-19-921846-2

10 9 8 7 6

Typeset by SPI Publisher Services, Pondicherry, India
Printed in Great Britain
on acid-free paper by
Ashford Colour Press Ltd, Gosport, Hampshire

To
Peter and Nina

Contents

List of illustrations

Chapter 1
The name of the game

What is game theory about?

When my wife was away for the day at a pleasant little conference in Tuscany, three young women invited me to share their table for lunch. As I sat down, one of them said in a sultry voice, 'Teach us how to play the game of love', but it turned out that all they wanted was advice on how to manage Italian boyfriends. I still think they were wrong to reject my strategic recommendations, but they were right on the nail in taking for granted that courting is one of the many different kinds of game we play in real life.

Drivers manoeuvring in heavy traffic are playing a driving game. Bargain-hunters bidding on eBay are playing an auctioning game. A firm and a union negotiating next year's wage are playing a bargaining game. When opposing candidates choose their platform in an election, they are playing a political game. The owner of a grocery store deciding today's price for corn flakes is playing an economic game. In brief, a game is being played whenever human beings interact.

Antony and Cleopatra played the courting game on a grand scale. Bill Gates made himself immensely rich by playing the computer software game. Adolf Hitler and Josef Stalin played a game that killed off a substantial fraction of the world's population. Kruschev and Kennedy played a game during the Cuban missile crisis that might have wiped us out altogether.

With such a wide field of application, game theory would be a universal panacea if it could always predict how people will play the many games of which social life largely consists. But game theory isn't able to solve all of the world's problems, because it only works when people play games *rationally*. So it can't predict the behaviour of love-sick teenagers like Romeo or Juliet, or madmen like Hitler or Stalin. However, people don't always behave irrationally, and so it isn't a waste of time to study what happens when people put on their thinking caps. Most of us at least try to spend our money sensibly – and we don't do too badly much of the time or economic theory wouldn't work at all.

Even when people haven't thought everything out in advance, it doesn't follow that they are necessarily behaving irrationally. Game theory has had some notable successes in explaining the behaviour of spiders and fish, neither of which can be said to think at all. Such mindless animals end up behaving as though they were rational, because rivals whose genes programmed them to behave irrationally are now extinct. Similarly, companies aren't always run by great intellects, but the market is often just as ruthless as Nature in eliminating the unfit from the scene.

Does game theory work?

In spite of its theoretical successes, practical men of business used to dismiss game theory as just one more ineffectual branch of social science, but they changed their minds more or less overnight after the American government decided to auction off the right to use various radio frequencies for use with cellular telephones.

With no established experts to get in the way, the advice of game theorists proved decisive in determining the design of the rules of the auctioning games that were used. The result was that the

American taxpayer made a profit of $20 billion – more than twice the orthodox prediction. Even more was made in a later British telecom auction for which I was responsible. We made a total of $35 billion in just one auction. In consequence, *Newsweek* magazine described me as the ruthless, Poker-playing economist who destroyed the telecom industry!

As it turned out, the telecom industry wasn't destroyed. Nor is it at all ruthless to make the fat cats of the telecom industry pay for their licences what they think they are worth – especially when the money is spent on hospitals for those who can't afford private medical care. As for Poker, it is at least 20 years since I played for more than nickels and dimes. The only thing that *Newsweek* got right is that game theory really does work when applied by people who know what they are doing. It works not just in economics, but also in evolutionary biology and political science. In my recent book *Natural Justice*, I even outrage orthodox moral philosophers by using game theory when talking about ethics.

Toy games

Each new big-money telecom auction needs to be tailored to the circumstances in which it is going to be run. One can't just take a design off the shelf, as the American government found when it hired Sotheby's to auction off a bunch of satellite transponders. But nor can one capture all the complicated ins and outs of a new telecom market in a mathematical model. Designing a telecom auction is therefore as much an art as a science. One extrapolates from simple models chosen to mimic what seem to be the essential strategic features of a problem.

I try to do the same in this book, which therefore contains no algebra and a minimum of technical jargon. It looks only at toy games, leaving aside all the bells and whistles with which they are complicated in real life. However, most people find that even toy games give them plenty to think about.

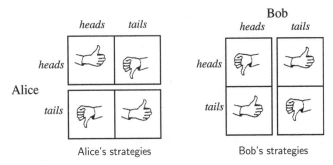

1. Alice and Bob's decision problem in Matching Pennies

Conflict and cooperation

Most of the games in this book have only two players, called Alice and Bob. The first game they will play is Matching Pennies.

Sherlock Holmes and the evil Professor Moriarty played Matching Pennies on the way to their final confrontation at the Reichenbach Falls. Holmes had to decide at which station to get off a train. Moriarty had to decide at which station to lie in wait. A real-life counterpart is played by dishonest accountants and their auditors. The former decide when to cheat and the latter decide when to inspect the books.

In our toy version, Alice and Bob each show a coin. Alice wins if both coins show the same face. Bob wins if they show different faces. Alice and Bob therefore each have two strategies, *heads* and *tails*. Figure 1 shows who wins and loses for all possible strategy combinations. These outcomes are the players' *payoffs* in the game. The thumbs-up and thumbs-down icons have been used to emphasize that payoffs needn't be measured in money.

Figure 2 shows how all the information in Figure 1 can be assembled into a payoff table, with Alice's payoff in the southwest corner of each cell, and Bob's in the northeast corner. It also shows a two-player version of the very different Driving Game that we

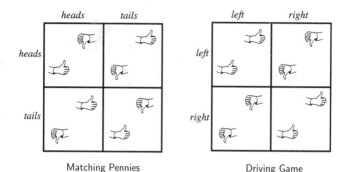

	heads	*tails*

Matching Pennies

	left	*right*

Driving Game

2. Payoff tables. Alice chooses a row and Bob chooses a column

play every morning when we get into our cars to drive to work. Alice and Bob again have two pure strategies, *left* and *right*, but now the players' payoffs are totally aligned instead of being diametrically opposed. When journalists talk about a win-win situation, they have something like the Driving Game in mind.

Von Neumann

The first result in game theory was John Von Neumann's minimax theorem, which applies only to games like Matching Pennies in which the players are modelled as implacable enemies. One sometimes still reads dismissive commentaries on game theory in which Von Neumann is caricatured as the archetypal cold warrior – the original for Dr Strangelove in the well known movie. We are then told that only a crazed military strategist would think of applying game theory in real life, because only a madman or a cyborg would make the mistake of supposing that the world is a game of pure conflict.

Von Neumann was an all-round genius. Inventing game theory was just a sideline for him. It is true that he was a hawk in the Cold War, but far from being a mad cyborg, he was a genial soul, who liked to party and have a good time. Just like you and me, he preferred cooperation to conflict, but he also understood that the way to achieve cooperation isn't to pretend that people can't sometimes profit by causing trouble.

Cooperation and conflict are two sides of the same coin, neither of which can be understood properly without taking account of the other. To consider a game of pure conflict like Matching Pennies isn't to claim that all human interaction is competitive. Nor is one claiming that all human interaction is cooperative when one looks at a game of pure coordination like the Driving Game. One is simply distinguishing two different aspects of human behaviour so that they can be studied one at a time.

Revealed preference

To cope with cooperation and conflict together, we need a better way of describing the motivation of the players than simply saying that they like winning and dislike losing. For this purpose, economists have invented the idea of *utility*, which allows each player to assign a numerical value to each possible outcome of a game.

In business, the bottom line is commonly profit, but economists know that human beings often have more complex aims than simply making as much money as they can. So we can't identify utility with money. A naive response is to substitute happiness for money. But what is happiness? How do we measure it?

It is unfortunate that the word 'utility' is linked historically with Victorian utilitarians like Jeremy Bentham and John Stuart Mill, because modern economists don't follow them in identifying utility with how much pleasure or how little pain a person may feel. The modern theory abandons any attempt to explain how people behave in terms of what is going on inside their heads. On the contrary, it makes a virtue of making no psychological assumptions at all.

We don't try to explain *why* Alice or Bob behave as they do. Instead of an explanatory theory, we have to be content with a descriptive theory, which can do no more than say that Alice or Bob will be acting inconsistently if they did such-and-such in the past, but now plan to do so-and-so in the future. In game theory, the object is to observe the decisions that Alice and Bob make (or would make) when they aren't interacting with each other or anyone else, and to deduce how they will behave when interacting in a game.

We therefore don't argue that some preferences are more rational than others. We follow the great philosopher David Hume in regarding reason as the 'slave of the passions'. As he extravagantly remarked, there would be nothing *irrational* about his preferring the destruction of the entire universe to scratching his finger. However, we go even further down this road by regarding reason purely as an instrument for avoiding inconsistent behaviour. Any consistent behaviour therefore counts as rational.

With some mild assumptions, acting consistently can be shown to be the same as behaving as though seeking to maximize the value of something. Whatever this abstract something may be in a particular context, economists call it utility. It needn't correlate with money, but it sadly often does.

Taking risks

In acting consistently, Alice may not be aware that she is behaving as though maximizing something we choose to call her utility. But if we want to predict her behaviour, we need to be able to measure her utility on a utility scale, much as temperature is measured on a thermometer. Just as the units on a thermometer are called degrees, we can then say that a *util* is a unit on Alice's utility scale.

The orthodoxy in economics used to be that such cardinal utility scales are intrinsically nonsensical, but Von Neumann fortunately didn't know this when Oskar Morgenstern turned up at his house one day complaining that they didn't have a proper basis for the numerical payoffs in the book on game theory they were writing together. So Von Neumann invented a theory on the spot that measures how much Alice wants something by the size of the risk she is willing to take to get it. We can then figure out what choice she will make in risky situations by finding the option that will give her the highest utility on average.

It is easy to use Von Neumann's theory to find how much utility to assign to anything Alice may need to evaluate. For example, how many utils should Alice assign to getting a date with Bob?

We first need to decide what utility scale to use. For this purpose, pick two outcomes that are respectively better and worse than any other outcome Alice is likely to encounter. These outcomes will correspond to the boiling and freezing points of water used to calibrate a Celsius thermometer, in that the utility scale to be constructed will assign 0 utils to the worst outcome, and 100 utils to the best outcome. Next consider a bunch of (free) lottery tickets in which the only prizes are either the best outcome or the worst outcome.

When we offer Alice lottery tickets with higher and higher probabilities of getting the best outcome as an alternative to a date with Bob, she will eventually switch from saying *no* to saying *yes*. If the probability of the best outcome on the lottery ticket that makes her switch is 75%, then Von Neumann's theory says that a date with Bob is worth 75 utils to her. Each extra percentage point added to her indifference probability therefore corresponds to one extra util.

When some people evaluate sums of money using this method, they always assign the same number of utils to each extra dollar. We call such people risk neutral. Those who assign fewer utils to each extra dollar than the one that went before are called risk averse.

Insurance

Alice is thinking of accepting an offer from Bob to insure her Beverley Hills mansion against fire. If she refuses his offer, she faces a lottery in which she ends up with her house plus the insurance premium if her house doesn't burn down, and with only the premium if it does. This has to be compared with her ending up for sure with the value of the house less the premium if she accepts Bob's offer.

If it is rational for Bob to make the offer and for Alice to accept, he must think that the lottery is better than breaking even for sure, and she must have the opposing preference. The existence of the insurance industry therefore confirms not only that it can be rational to gamble – provided that the risks you take are calculated risks – but that rational people can have different attitudes to taking risks. In the insurance industry, the insurers are close to being risk neutral and the insurees are risk averse to varying degrees.

Notice that economists regard the degree of risk aversion that a person reveals as a matter of personal preference. Just as Alice may or may not prefer chocolate ice-cream to vanilla, so she may or may not prefer to spend $1,000 on insuring her house. Some philosophers – notably John Rawls – insist that it is *rational* to be risk averse when defending whatever alternative to maximizing average utility they prefer, but such appeals miss the point that the players' attitudes to taking risks have already been taken into account when using Von Neumann's method to assign utilities to each outcome.

Economists make a different mistake when they attribute risk aversion to a dislike of the act of gambling. Von Neumann's theory only makes sense when the players are entirely neutral to the actual act of gambling. Like a Presbyterian minister insuring his house, they don't gamble because they enjoy gambling – they gamble only when they judge that the odds are in their favour.

	heads	*tails*
heads	-1 / +1	+1 / -1
tails	+1 / -1	-1 / +1

Matching Pennies

	left	*right*
left	+1 / +1	-1 / -1
right	-1 / -1	+1 / +1

Driving Game

3. Numerical payoffs

Life isn't a zero-sum game

As with measuring temperature, we are free to choose the zero and the unit on Alice's utility scale however we like. We could, for example, have assigned 32 utils to the worst outcome, and 212 utils to the best outcome. The number of utils a date with Bob is worth on this new scale is found in the same way that one converts degrees Celsius into degrees Fahrenheit. So the date with Bob that was worth 75 utils on the old scale would be worth 167 utils on the new scale.

In the toy games we have considered so far, Alice and Bob have only the outcomes WIN and LOSE to evaluate. We are free to assign these two outcomes any number of utils we like, as long as we assign more utils to winning than to losing. If we assign plus one util to winning and minus one util to losing, we get the payoff tables of Figure 3.

The payoffs in each cell of Matching Pennies in Figure 3 always add up to zero. We can always fix things to make this true in a game of pure conflict. Such games are therefore said to be *zero sum*. When gurus tell us that life isn't a zero-sum game, they therefore aren't saying anything about the total sum of happiness in the world. They are just reminding us that the games we play in real life are seldom games of pure conflict.

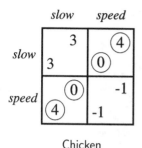

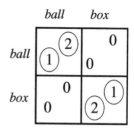

Chicken

Battle of the Sexes

4. Games with mixed motivations

Nash equilibrium

The old movie *Rebel without a Cause* still occasionally gets a showing because it stars the unforgettable James Dean as a sexy teenage rebel. The game of Chicken was invented to commemorate a scene in which he and another boy drive cars towards a cliff edge to see who will chicken out first. Bertrand Russell famously used the episode as a metaphor for the Cold War.

I prefer to illustrate Chicken with a more humdrum story in which Alice and Bob are two middle-aged drivers approaching each other in a street too narrow for them to pass safely without someone slowing down. The strategies in Figure 4 are therefore taken to be *slow* and *speed*.

The new setting downplays the competitive element of the original story. Chicken differs from zero-sum games like Matching Pennies because the players also have a joint interest in avoiding a mutual disaster.

The stereotypes embedded in the Battle of the Sexes pre-date the female liberation movement. Alice and Bob are a newly married couple honeymooning in New York. At breakfast, they discuss whether to go to a boxing match or the ballet in the evening, but

5. James Dean

fail to make a decision. They later get separated in the crowds and now each has to decide independently where to go in the evening.

The story that accompanies the Battle of the Sexes emphasizes the cooperative features of their problem, but there is also a conflictual element absent from the Driving Game, because each player prefers that they coordinate on a different outcome. Alice prefers the ballet and Bob the boxing match.

John Nash

Everybody has heard of John Nash now that his life has been featured in the movie *A Beautiful Mind*. As the movie documents, the highs and lows of his life are out of the range of experience of most human beings. He was still an undergraduate when he initiated the modern theory of rational bargaining. His graduate thesis formulated the concept of a Nash equilibrium, which is now regarded as the basic building block of the theory of games. He went on to solve major problems in pure mathematics, using methods of such originality that his reputation as a mathematical genius of the first rank became firmly established. But he fell prey

6. John Nash

to a schizophrenic illness that wrecked his career and finally left him to languish in obscurity for more than 40 years as an object of occasional mockery on the Princeton campus. His recovery in time to be awarded a Nobel Prize in 1994 seems almost miraculous in retrospect. But as Nash comments, without his 'madness', he would perhaps only have been another of the faceless multitudes who have lived and died on this planet without leaving any trace of their existence behind.

However, one doesn't need to be a wayward genius to understand the idea of a Nash equilibrium. We have seen that the payoffs in a game are chosen to make it tautological that rational players will seek to maximize their average payoff. This would be easy if players knew what strategies their opponents were going to choose. For example, if Alice knew that Bob were going to choose *ball* in the Battle of the Sexes, she would maximize her payoff by choosing *ball* as well. That is to say, *ball* is Alice's best reply to Bob's choice of *ball*, a fact indicated in Figure 4 by circling Alice's payoff in the cell that results if both players choose *ball*.

A Nash equilibrium is just a pair of strategies whose use results in a cell in which *both* payoffs are circled. More generally, a Nash equilibrium occurs when all the players are simultaneously making a best reply to the strategy choices of the others.

Both (*box*, *box*) and (*ball*, *ball*) are therefore Nash equilibria in the Battle of the Sexes. Similarly, (*slow*, *speed*) and (*speed*, *slow*) are Nash equilibria in Chicken.

Why should we care about Nash equilibria? There are two major reasons. The first supposes that ideally rational players reason their way to a solution of a game. The second supposes that people find their way to a solution by some evolutionary process of trial and error. Much of the predictive power of game theory arises from the possibility of passing back and forth between these alternative interpretations. We seldom know much about the details of evolutionary processes, but we can sometimes leap ahead to predict where they will eventually end up by asking what rational players would do in the situation under study.

Rational interpretation

Suppose that somebody even cleverer than Nash or Von Neumann had written a book that lists all possible games along with an authoritative recommendation on how each game should be played by rational players. Such a great book of game theory would necessarily have to pick a Nash equilibrium as the solution of each game. Otherwise it would be rational for at least one player to deviate from the book's advice, which would then fail to be authoritative.

Suppose, for example, that the book recommended that teenage boys playing Chicken should both choose *slow* as their mothers would wish. If the book were authoritative, each player would then know that the other was going to play *slow*. But a rational player in Chicken who knows that his opponent is going to choose *slow* will necessarily choose *speed*, thereby refuting the book's claim to be authoritative.

Notice that the reasoning in this defence of Nash equilibria is circular. Why does Alice play this way? Because Bob plays that way. Why does Bob play that way? Because Alice plays this way.

Various Latin tags are available to those who are unhappy with such circular arguments. When first accused of committing the fallacy of *circulus in probando* when talking about equilibria, I had to go and look it up. It turns out that I was lucky not to have been accused of the even more discreditable *petitio principii*. But all arguments must obviously either be circular or reduce to an infinite regression if one never stops asking *why*. Dictionary definitions are the most familiar example.

In games, we can either forever contemplate the infinite regression that begins:

Alice thinks that Bob thinks that Alice thinks that Bob thinks . . .

or else take refuge in the circularity built into the idea of a Nash equilibrium. This short circuits the infinite regression by observing that any other strategy profile will eventually be destabilized when the players start thinking about what the other players are thinking. Or to say the same thing another way, if the players' beliefs about each other's plans are to be consistent, then they must be in equilibrium.

Evolutionary interpretation

The rational interpretation of Nash equilibrium had such a grip on early game theorists that the evolutionary interpretation was almost entirely neglected. The editors of the journal in which Nash published his paper on equilibria even threw out his remarks on this subject as being without interest! But game theory would never be able to predict the behaviour of ordinary people if the evolutionary interpretation were invalid. For example, the famous mathematician Emile Borel thought about game theory before Von Neumann but came to the conclusion that the minimax theorem was probably false. So what hope would there be for the rest of us, if even someone as clever as Borel couldn't reason his way to a solution of the simplest class of games!

There are many possible evolutionary interpretations of Nash equilibria, which differ in the adjustment process by means of which players may find their way to an equilibrium. In the simpler adjustment processes, the payoffs in a game are identified with how fit the players are. Processes that favour fitter strategies at the expense of their less successful brethren can then only stop working when we get to a Nash equilibrium, because only then will all the surviving strategies be as fit as it is possible to be in the circumstances. We therefore don't need our players to be mathematical whizzes for Nash equilibria to be relevant. They often predict the behaviour of animals quite well. Nor is the evolutionary significance of Nash equilibria confined to biology. They have a predictive role whenever an adjustment process tends to eliminate strategies that generate low payoffs.

For example, stockbrokers who do less well than their competitors go bust. The rules-of-thumb that stockbrokers use are therefore subject to the same kind of evolutionary pressures as the genes of fish or insects. It therefore makes sense to look at Nash equilibria in the games played by stockbrokers, even though we all know that some stockbrokers wouldn't be able to find their way around a goldfish bowl, let alone a game theory book.

Prisoner's Dilemma

The most famous toy game of all is the Prisoner's Dilemma. In the traditional story used to motivate the game, Alice and Bob are gangsters in the Chicago of the 1920s. The District Attorney knows that they are guilty of a major crime, but is unable to convict either unless one of them confesses. He orders their arrest, and separately offers each the following deal:

If you confess and your accomplice fails to confess, then you go free. If you fail to confess but your accomplice confesses, then you will be convicted and sentenced to the maximum term in jail. If you both confess, then you will both be convicted, but the maximum sentence will not be imposed. If neither confesses, you will both be framed on a tax evasion charge for which a conviction is certain.

The story becomes more poignant if Alice and Bob have agreed to keep their mouths shut if ever put into such a situation. Holding out then corresponds to cooperating and confessing to defecting, as in the table on the left of Figure 7. The payoffs in the table correspond to notional years in jail (on the assumption that one util always corresponds to one extra year of freedom).

A less baroque story assumes that Alice and Bob each have access to a pot of money. Both are independently allowed either to give their opponent $2 from the pot, or to put $1 from the pot in their own pocket. On the assumption that Alice and Bob care only about money, we are led to the payoff table on the right of Figure 7 in which utils have been identified with dollars. In this case, the altruistic strategy of giving $2 has been assigned the label *dove*,

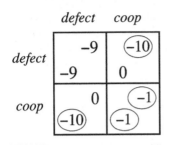

	defect	coop
defect	-9 / -9	(-10) / 0
coop	0 / (-10)	(-1) / (-1)

gangster version

	dove	hawk
dove	2 / 2	(3) / 0
hawk	0 / (3)	(1) / (1)

give-or-take version

7. **Two versions of the Prisoner's Dilemma: in the version on the right, *dove* represents giving and *hawk* represents taking**

and the selfish strategy of taking $1 has been assigned the label *hawk*.

Circling best replies reveals that the only Nash equilibrium in the give-or-take version of the Prisoner's Dilemma is for both Alice and Bob to play *hawk*, although each would get more if they both played *dove*. The gangster version is strategically identical. In the unique Nash equilibrium, each will defect, with the result that they will both spend a long time in jail, although each would get a much lighter sentence if they both cooperated.

Paradox of rationality?

A whole generation of scholars swallowed the line that the Prisoner's Dilemma embodies the essence of the problem of human cooperation. They therefore set themselves the hopeless task of giving reasons why game theory's resolution of this supposed 'paradox of rationality' is mistaken (See Fallacies of the Prisoner's Dilemma, Chapter 10). But game theorists think it just plain wrong that the Prisoner's Dilemma captures what matters about human cooperation. On the contrary, it represents a situation in which the dice are as loaded against the emergence of cooperation as they could possibly be.

If the great game of life played by the human species were adequately modelled by the Prisoner's Dilemma, we wouldn't have evolved as social animals! We therefore see no more need to solve an invented paradox of rationality than to explain why people drown when thrown into Lake Michigan with their feet encased in concrete. No paradox of rationality exists. Rational players don't cooperate in the Prisoner's Dilemma because the conditions necessary for rational cooperation are absent.

Fortunately the paradox-of-rationality phase in the history of game theory is just about over. Insofar as they are remembered, the many fallacies that were invented in hopeless attempts to show that it is rational to cooperate in the Prisoner's Dilemma are now mostly quoted as entertaining examples of what psychologists call magical reasoning, in which logic is twisted to secure some desired outcome. My favourite example is Immanuel Kant's claim that rationality demands obeying his categorical imperative. In the Prisoner's Dilemma, rational players would then all choose *dove*, because this is the strategy that would be best if everybody chose it.

Domination

The idea that it is necessarily irrational to do things that would be bad if everybody did them is very pervasive. Your mother was probably as fond of this argument as mine. The following knock-down refutation in the case of the Prisoner's Dilemma is therefore worth repeating.

So as not to beg any questions, we begin by asking where the payoffs that represent the players' preferences in the Prisoner's Dilemma come from. The theory of revealed preference tells us to find the answer by observing the choices that Alice and Bob make (or would make) when solving one-person decision problems.

Writing a larger payoff for Alice in the bottom-left cell of the payoff table of the Prisoner's Dilemma than in the top-left cell therefore means that Alice would choose *hawk* in the one-person decision problem that she would face if she knew in advance that Bob had chosen *dove*. Similarly, writing a larger payoff in the bottom-right cell means that Alice would choose *hawk* when faced with the one-person decision problem in which she knew in advance that Bob had chosen *hawk*.

The very definition of the game therefore says that *hawk* is Alice's best reply when she knows that Bob's choice is *dove*, and also when she knows his choice is *hawk*. So she doesn't need to know anything about Bob's actual choice to know her best reply to it. It is rational for her to play *hawk* whatever strategy he is planning to choose. In this unusual circumstance, we say that *hawk* dominates Alice's alternative strategies.

Objections?

Two objections to the preceding analysis are common. The first denies that Alice would choose to defect in the gangster version of the Prisoner's Dilemma if she knew that Bob had chosen to cooperate. Various reasons are offered that depend on what one believes about conditions in Al Capone's Chicago, but such objections miss the point. If Alice wouldn't defect if she knew that Bob had chosen to cooperate, then she wouldn't be playing the Prisoner's Dilemma. Here and elsewhere, it is important not to take the stories used to motivate games too seriously. It is the payoff tables of Figure 7 that define the Prisoner's Dilemma – not the silly stories that accompany them.

The second objection always puzzles me. It is said that appealing to the theory of revealed preference reduces the claim that it is rational to defect in the Prisoner's Dilemma to a tautology. Since tautologies have no substantive content, the claim can therefore be ignored! But who would say the same of $2 + 2 = 4$?

Experiments

An alternative response is to argue that it doesn't matter what is rational in the Prisoner's Dilemma, because laboratory experiments show that real people actually play *dove*. The payoffs in such experiments aren't usually determined using the theory of revealed preference. They are nearly always just money, but the results can nevertheless be very instructive.

Inexperienced subjects do indeed cooperate a little more than half the time on average, but the evidence is overwhelming in games like the Prisoner's Dilemma that the rate of defection increases steadily as the subjects gain experience, until only about 10% of subjects are still cooperating after ten trials or so.

Computer simulations are also mentioned which supposedly show that evolution will eventually generate cooperation in the Prisoner's Dilemma, but such critics have usually confused the Prisoner's Dilemma with its indefinitely repeated cousin in which cooperation is indeed a Nash equilibrium (See Tit-for-tat, Chapter 5).

Chapter 2
Chance

Conan Doyle's analysis of his version of Matching Pennies in *The Final Problem* doesn't reflect much credit on his hero's supposed intellectual mastery. Edgar Allan Poe does better in the *Purloined Letter*, in which the villain has stolen a letter, and the problem is where to look for it.

Poe argues that the way to win is to extend chains of reasoning of the form 'He thinks that I think that he thinks that I think ...' one step further than your opponent. In defence of this proposition, he invents a boy who consistently wins at Matching Pennies by imitating his opponent's facial expression, thereby supposedly learning what he must be thinking. It is admittedly amazing how many Poker players give their hands away by being unable to control their body language, but Alice and Bob can't *both* use Poe's trick successfully even if neither ever learns to keep a Poker face.

Game theory escapes the apparent infinite regression with which Alice and Bob are faced by appealing to the idea of a Nash equilibrium. But we are still left with a problem, because the trick of circling best replies doesn't work for Matching Pennies. After circling all the payoffs in Figure 3 that are best replies, we end up with two Nash equilibria in the Driving Game, but none at all in Matching Pennies.

This fact may seem mysterious to those who remember that John Nash won his Nobel Prize partly for showing that all finite games have at least one equilibrium. The answer to the mystery is that we need to look beyond the pure strategies we have considered up to now, and consider mixed strategies as well.

Does randomizing make sense?

A mixed strategy requires that players randomize their choice of pure strategy. It is natural to object that only crazy people make serious decisions at random, but mixed strategies are used all the time without anyone realizing it.

My favourite example arose when I was advising a package holiday company on a regulatory matter. Game theory predicts that such a company will use a mixed strategy in the pricing game it has to play when the demand for vacations proves to be unexpectedly low. However, when I asked a senior executive whether his company actively randomized their prices last year, he reacted with horror at such an outlandish suggestion. So why were his prices for similar vacations so very different? His answer was instructive: 'You have to keep the opposition guessing.'

His answer shows that he understood perfectly well *why* game theory sometimes recommends the use of mixed strategies. What he didn't want to face up to is that his company's method for setting prices was essentially a randomizing device. Nobody cut any cards. Nobody rattled a dice box. But from the point of view of a rival trying to predict what his company would charge for two weeks in the Bahamas, they might as well have done so.

Mixed Nash equilibria

The use of mixed strategies isn't at all surprising in Matching Pennies, where the whole point is to keep the opponent guessing. As every child knows, the solution is to randomize between *heads*

8. Rolling dice

and *tails*. If both players use this mixed strategy, the result is a Nash equilibrium. Each player wins half the time, which is the best that both can do given the strategy choice of the other.

Similarly, it is a Nash equilibrium in the Driving Game if both players choose *left* and *right* with equal probability, which therefore has three Nash equilibria, two pure and one mixed. The same is also true in both Chicken and the Battle of the Sexes, but the mixed Nash equilibrium in the Battle of the Sexes requires more of the players than that they simply make each of their pure strategies equally likely.

In the Battle of the Sexes, Bob likes boxing twice as much as ballet, and so Alice must play *box* half as often as *ball* to ensure that he gets the same payoff on average from his two pure strategies. Since Bob doesn't then care which of his pure strategies gets played, all of his strategies are then equally good – including the mixed strategy which makes *ball* half as likely as *box*. But the use of this mixed strategy makes Alice indifferent between her two pure strategies. So all of her strategies are then equally good – including the mixed strategy which makes *box* twice as likely as *ball*. This completion of the circuit shows that we have found a mixed Nash equilibrium in which Alice and Bob each play their more favoured strategy two-thirds of the time.

Making the other guy indifferent

Rational players never randomize between two pure strategies unless they are indifferent between them. If one strategy were better, the inferior strategy would never get played at all. What might make you indifferent between two strategies? In the Battle of the Sexes, the reason is that you believe your opponent is going to play a mixed strategy that equalizes the average payoff you get from each of your strategies. This feature of a mixed Nash equilibrium sometimes leads to results that look paradoxical at first sight.

The Good Samaritan Game is played by a whole population of identical players, all of whom want someone to respond to a cry for help. Each player gets ten utils if someone helps, and nothing if nobody helps. The snag is that helping is a nuisance, and so all the players who offer help must subtract one util from their payoffs.

If nobody else is planning to help, you do best by offering to help yourself. If everybody else is planning to help, you maximize your payoff by doing nothing. So the only possible Nash equilibrium in which everybody independently uses the same strategy is necessarily mixed. In such a mixed Nash equilibrium, there must be precisely one chance in ten that nobody else offers help, because this is the frequency that makes you indifferent between helping and not helping.

The actual probability that help is offered in equilibrium is somewhat higher, because there is some chance that you will offer to help yourself. However, the probability that any single player offers help in equilibrium has got to get smaller as the population gets larger because the probability that nobody else helps has to stay equal to 1/10. So the bigger the population, the lower the chances that anyone will help. With only two players, each helps with probability 9/10 and the cry for help is ignored only one time in a hundred. With a million players, each helps with such a tiny probability that nobody at all answers the cry for help about one time in ten.

The consequences can be chilling, as a notorious case in New York illustrates. A woman was assaulted at length after dark, and finally murdered in the street. Many people heard her cries for help but nobody even phoned the police. Should we follow the newspapers and deduce that city life makes monsters of us all? Perhaps it does, but the Good Samaritan Game suggests that even small-town folk might behave in the same way if put in the same situation.

Voting has a similar character. To take an extreme case, suppose that Alice and Bob are the only candidates for the presidency. It is common knowledge that Bob is a hopeless case; only his mother thinks he would be the better president. She is sure to vote, but why should anyone else bother? As in the Good Samaritan Game, adding more voters makes things worse. In equilibrium, Bob will get elected with some irreducible probability even if there are a million voters.

Such voting games are only toys. Real people seldom think rational thoughts about whether or not to vote. Even if they did, they might feel that going to the polling booth is a pleasure rather than a pain. But the model nevertheless shows that the pundits who denounce the large minority of people who fail to vote in presidential elections as irrational are talking through their hats. If we want more people to vote, we need to move to a more decentralized system in which every vote really does count enough to outweigh the lack of enthusiasm for voting which so many people obviously feel. If we can't persuade such folk that they like to vote and we don't want to change our political system, we will just have to put up with their staying at home on election night. Simply repeating the slogan that 'every vote counts' isn't ever going to work, because it isn't true.

Getting to equilibrium

How do people find their way to a Nash equilibrium? This question is particularly pressing in the case of mixed equilibria. Why should Alice adjust her behaviour to make Bob indifferent between some of his strategies?

Sports studies show that athletes sometimes behave in quite close accord with game theory predictions. Taking penalty kicks in soccer is one example. Where should the ball be aimed? Which way should the goalkeeper jump? Tennis is another example. Should I smash or should I lob? It seems unlikely that coaches read any game theory books, so how come they know the correct frequency with which to choose each option? Presumably they learn by trial and error.

Nobody understands all the different ways in which real people learn new ways of doing things, but we have some toy models that capture some of what must be going on. Even the following naive model does surprisingly well.

Chance

Alice and Bob are robots who play the same game repeatedly. At each repetition, Alice is programmed to play her best reply to a mixed strategy in which each of Bob's pure strategies is played with the same frequency he has played it in the past. Bob has the same program, so neither he nor Alice are fully rational, because they could both sometimes improve their payoffs if they were programmed more cleverly. Game theorists say that they are only boundedly rational.

As time passes, the frequencies with which the robots have played their second pure strategy evolve as shown in Figure 9 (which has been simplified by passing from discrete to continuous time). For example, Alice's best reply in Matching Pennies is *tails* whenever the current frequency with which Bob has played *tails* exceeds one half. So her frequency for *tails* will increase until his frequency for

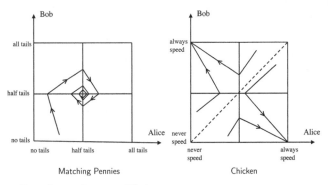

      Chicken

9. Learning to play an equilibrium

tails falls below one half, after which it will abruptly begin to decrease.

Following the arrows in Figure 9 always leads to a Nash equilibrium. No matter how we initialize the robots, someone counting how often they play each of their pure strategies will therefore eventually find it hard to distinguish one of our boundedly rational robots from a perfectly rational player.

In the case of Matching Pennies, which is closest to tennis or soccer, the frequencies with which *heads* or *tails* are played always converge on their equilibrium values of $1/2$. In laboratory experiments with human subjects, the general pattern is much the same, although the frequencies don't evolve in such a regular manner and they begin to drift when they get near enough to a mixed equilibrium, because the players are then nearly indifferent between the available strategies.

The situation in Chicken is more complicated. Each pure equilibrium has a basin of attraction. If we initialize our robots so that they begin in the basin of attraction of a particular equilibrium, they will eventually converge on that equilibrium. The basin of attraction for (*slow*, *speed*) lies above the diagonal in

Figure 9. The basin of attraction for (*speed*, *slow*) lies below the diagonal. The basin of attraction for the mixed equilibrium is just the diagonal itself.

It is easy to construct games in which the behaviour of robots like Alice and Bob would cycle forever without ever settling down on an equilibrium, but human beings are capable of learning in more sophisticated ways than Alice or Bob. In particular, we commonly enjoy a great deal of feedback from all kinds of sources when learning how to behave when faced with a new game.

For example, rookie stockbrokers learn the ropes from their more experienced colleagues. Young scientists peruse the history of Nobel laureates in the hope of finding the secret of their success. Novelists tediously recycle the plots of the latest best-seller. Shoppers tell each other where the best bargains are to be found. Toy models of such social or imitative learning converge more quickly and reliably on Nash equilibria than models in which single individuals learn by trial and error.

Evolutionary game theory is the study of such dynamic models. Its application to evolutionary biology is so important that it gets a chapter all to itself (Chapter 8).

Minimax theorem

When a youthful John Nash called at Von Neumann's office to tell him of his proof that all finite games have at least one equilibrium when mixed strategies are allowed, Von Neumann was dismissive. Why didn't he welcome Nash's contribution?

It is true that the method Nash used to prove his theorem wasn't anything new for Von Neumann, who had pioneered the method himself. It is also true that Nash's approach was probably not very tactful, since he famously called on Albert Einstein around the same time to tell him how to do physics. But Von Neumann had nothing to fear from a brash young graduate student muscling in on his domain. I think there was a more fundamental reason for Von Neumann's lack of interest.

Von Neumann never seems to have thought much about the evolutionary interpretation of game theory. He believed that the purpose of studying a game should be to identify an unambiguous rational solution. The idea of a Nash equilibrium doesn't meet this requirement, because most games have many Nash equilibria, and there is often no purely rational reason for selecting one equilibrium rather than another. As Von Neumann later remarked, the best-reply criterion only tells us that some strategy profiles *can't* be the rational solution to a game, but we want to know which strategy profiles *can* be regarded as solutions.

Minimax and maximin

Von Neumann presumably restricted his attention to two-person, zero-sum games because they are one of the few classes of games in which his ideal of a unique rational solution can be realized. It is unfortunate that his proof of this fact should be called the *minimax* theorem, because the rational solution of a two-person, zero-sum game is actually for each player to apply the *maximin* principle. This tells you to work out the worst payoff you could get on average from each of your mixed strategies, and then to choose whichever strategy would maximize your payoff if this worst-case scenario were always realized.

For example, the worst thing that could happen to Alice in Matching Pennies is that Bob will guess her choice of mixed strategy. If this mixed strategy requires her to play *heads* more than half the time, then he will play *tails* all the time. She will then lose more than half the time and so her payoff will be negative. If Alice's mixed strategy requires her to play *tails* more than half the time, then Bob will play *heads* all the time. She will again lose more than half the time and so her payoff will again be negative.

Alice's maximin strategy is therefore to play *heads* and *tails* equally often, which guarantees her a payoff of exactly zero.

Only a paranoic would find the maximin principle attractive in general, since it assumes that the universe has singled you out to be its personal enemy. However, if Alice is playing Bob in a zero-sum game, he is the relevant universe and so the universe is indeed her personal enemy in this special case.

Why maximin?

Ironically, Von Neumann's minimax theorem follows immediately from Nash's proof that all finite games have at least one Nash equilibrium.

To see this, begin by locating a Nash equilibrium in a two-person, zero-sum game. Call Alice's equilibrium strategy *row* and Bob's equilibrium strategy *column*. The equilibrium payoffs will be called *Alice's value* and *Bob's value*. For example, in Matching Pennies both *row* and *column* are the mixed strategy in which *heads* and *tails* are played with equal probability; *Alice's value* and *Bob's value* are the zero payoff that each player gets on average if they play this way.

Alice can't be sure of getting more than *Alice's value* because Bob might always play *column,* to which her best reply is *row.* On the other hand, Alice can be sure of getting at least *Alice's value* by playing *row* because the best that Bob can do is to reply with *column,* and the best that Bob can do for himself in a zero-sum game is the same as the worst he can do to Alice. So *Alice's value* is Alice's maximin payoff, and *row* is one of her maximin strategies.

By the same reasoning, *Bob's value* is his maximin payoff and *column* is one of his maximin strategies. Since if *Alice's value* and *Bob's value* sum to zero, it follows that so do their maximin payoffs. Neither player can therefore get more than his or her maximin payoff unless the other gets less. So one can't improve on the maximin principle when playing a two-person, zero-sum game against a rational opponent.

Von Neumann's proof of this fact is called the minimax theorem, because saying that Alice and Bob's maximin payoffs sum to zero is equivalent to saying that Alice's maximin payoff equals her minimax payoff. But one mustn't make the common mistake of thinking that Von Neumann therefore recommended using the minimax principle. Nobody would want to work out the *best* payoff you could get on average from each of your mixed strategies, and then choose whichever strategy would *minimize* your payoff if this best-case scenario were always realized!

Finding maximin strategies

In retrospect, it is a pity that mathematicians took an immediate interest in the minimax theorem. The study of pursuit-evasion games in which a pilot seeks to evade a heat-seeking missile is certainly an interesting exercise in control theory, but such work naturally reinforces the prejudices of critics who are fixated on the idea that game theorists are mad cyborgs. Nor is the popularity of game theory likely to be enhanced by the abstruse finding that the minimax theorem can only be true in certain infinite games if we are willing to deny the Axiom of Choice. Game theory would have found a more ready acceptance in its early years if enthusiasts hadn't made it all seem so difficult.

Rock-Scissors-Paper

Every child knows this game. Alice and Bob simultaneously make a hand signal that represents one of their three pure strategies: *rock, scissors, paper*. The winner is determined by the rules:

rock	blunts	*scissors*
scissors	cuts	*paper*
paper	wraps	*rock* .

If both players make the same signal, the result is a draw, which both players regard as being equivalent to a lottery in which they win or lose with equal probability, so that the game is zero-sum.

It is obvious that the rational solution is for each player to use each of their three pure strategies equally often. They each then guarantee their maximin payoff of zero. The chief interest of the game is that one has to work very hard to find an evolutionary process that converges on this solution.

For example, the best-reply dynamics of Figure 9 end up cycling in a manner that periodically nearly eliminates each strategy in turn. One might dismiss this outcome as a curiosity if it weren't for the fact that the population mix of three varieties of Central American salamanders who play a game like Rock-Scissors-Paper also end up in a similar cycle, so that one variety always seems on the edge of extinction.

O'Neill's Card Game

Barry O'Neill used this game in the first laboratory experiment that found positive support for the maximin principle. Previous experiments had been discouraging. The eminent psychologist William Estes was particularly scathing when reporting on his test of Von Neumann's theory: 'Game theory will be no substitute for an empirically grounded behavioral theory when we want to predict what people will actually do in competitive situations.'

But in the experiment on which Estes based his dismissive remarks, there were only two subjects, who are described as being well practised in the reinforcement learning experiments that Estes was using to defend the (now discredited) theory of 'probability matching'. Neither subject knew that they were playing a game with another person. Even if they had known they were playing a game, the minimax theory would have been irrelevant to their plight, since they weren't told in advance what the payoffs of the game were. They were therefore playing with incomplete information – a situation to which Von Neumann's minimax theory doesn't apply.

In designing an experiment without such errors, O'Neill wanted to control for the possibility that subjects might have different attitudes to taking risks. For example, Rock-Scissors-Paper wouldn't be zero-sum if Alice and Bob didn't both think a draw is equivalent to winning or losing with equal probability. So O'Neill experimented on a game with only winning or losing, but which still has enough structure to make the solution unobvious.

Alice and Bob each have the ace and the picture cards from one of the suits in a deck of playing cards. They simultaneously show a card. Alice wins if both show an ace, or if there is a mismatch of picture cards. Otherwise Bob wins.

Alice's maximin strategy is found by asking which of her mixed strategies makes Bob indifferent between all his pure strategies. The answer is that Alice should play each picture card equally often and her ace twice as often. Bob should do the same, with the result that Alice will win two-fifths of the time and Bob will win three-fifths of the time.

Duel

The game of Duel is the nearest we are going to get to a military application. Alice and Bob walk towards each other armed with pistols loaded with just one bullet. The probability of either hitting the other increases the nearer the two approach. The payoff to each player is the probability of surviving.

How close should Alice get to Bob before firing? This is literally a question of life and death because, if she fires and misses, Bob will be able to advance to point-blank range with fatal consequences for Alice. Since someone dies in each possible outcome of the game, the payoffs therefore always sum to one.

One conclusion is obvious. It can't be a Nash equilibrium for one player to plan to fire sooner than the other, because it would be a better reply for the player who is planning to fire first to wait a tiny bit longer. But how close will they be when they simultaneously open fire?

The minimax theorem gives the answer right away. Duel is unit-sum rather than zero-sum, but the minimax theorem still applies (provided the payoffs still sum to one when the players fire simultaneously). The only difference is that the players' maximin payoffs now add up to one instead of zero. So if Alice is always twice as likely to hit Bob as he is to hit her, they will both fire at whatever distance makes Alice hit Bob two-thirds of the time and Bob hit Alice one-third of the time.

Chapter 3
Time

Games with perfect information

People sometimes think it frivolous to talk about human social problems as though they were mere parlour games. The advantage is that nearly everybody is able to think dispassionately about the strategic issues that arise in games like Chess or Poker, without automatically rejecting a conclusion if it turns out to be unwelcome. But logic is the same wherever it is applied.

Parlour games

At first sight, it doesn't look like Chess and Poker can be represented by payoff tables, because time enters the picture. It not only matters who does what – it matters when they do it.

Some of the difference is illusory. In the general case, a pure strategy is a plan of action that tells a player what to do under all possible contingencies that might arise in a game. The players can then be envisaged as choosing a strategy once and for all at the beginning of the game, and then delegating the play of the game to a robot. The resulting *strategic form* of Chess will then look just like Chicken or the Battle of the Sexes, except that its payoff table will be zero sum and have immensely more rows and columns.

Von Neumann argued that the first thing one should do in any game is to reduce it to its strategic form, which he called its normal form for this reason. However, the case of Chess makes it clear that this isn't always a very practical proposal, since it has more pure strategies than the estimated number of electrons in the known universe! Even when the strategic form isn't hopelessly unwieldy, it is often a lot easier to work things out by sticking with the *extensive form* of the game.

Game theorists use the analogy of a tree when describing a game in extensive form. Each move corresponds to a point called a node where the tree branches. The root of the tree corresponds to the first move of the game. The branches at each node correspond to the choices that can be made at that move. The leaves of the tree correspond to the final outcomes of the game, and so we must say who gets what payoff at each leaf. We must also say which player moves at each node, and what that player knows about what has happened so far in the game when making the move.

In Poker, the first move is made by a fictional player called Chance who shuffles and deals hands to the real players. What the players know about this move is extremely important in Poker, since the game would be devoid of interest if everybody knew what everybody else had been dealt. However, such games of imperfect information will be left until the next chapter. All the games in this chapter will be games of perfect information, in which nothing that has happened in the game so far is hidden from players when they make a move. Nor shall we consider games of perfect information like Duel that have chance moves. Chess is therefore the archetypal example for this chapter.

Backward induction

Backward induction is a contentious topic, but everybody agrees that we would always be able to use it to find the players' maximin values in a finite game of perfect information – if we had a large enough computer and sufficient time. Given a large enough lever and a place on which to stand, Archimedes was similarly correct when he said he would be able to move the world. Applying backward induction to Chess illustrates both its theoretical virtues and its practical drawbacks.

Chess

Label each leaf of the game tree for Chess with WIN, LOSE, or DRAW, depending on the outcome for White. Now pick any penultimate node (where each choice leads immediately to a leaf of the tree). Find the best choice for the player who moves at this node. Label the penultimate node with the label of the leaf to which this choice leads. Finally, throw away all of the tree that follows the penultimate node, which now becomes a leaf of a smaller tree in which the players' maximin values are unchanged.

Now do the same again and again, until all that is left is a label attached to the root of the original tree. This label is White's maximin outcome.

No matter how big and fast the computers we eventually build, they will never be able to complete this program for Chess, because it would take too long. So we will probably never know the solution of Chess. But at least we have established that, unlike Bigfoot or the Loch Ness Monster, there really is a solution for Chess.

If White's maximin outcome is WIN, then she has a pure strategy that guarantees her a victory against any defence by Black. If White's maximin outcome is LOSE, Black has a pure strategy that guarantees him a victory against any defence by White. However, most experts guess that White's maximin outcome is DRAW, which implies that both White and Black have pure strategies that guarantee a draw against any defence.

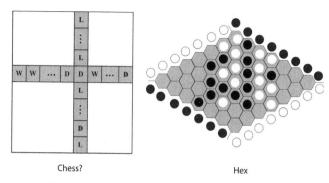

Chess?   Hex

10. Two board games

If these experts are right, then the strategic form of Chess has a
row in which all the outcomes are WIN or DRAW and a column in
which all the payoffs are LOSE or DRAW as in Figure 10. Without
the backward induction argument, I am not sure that this fact
would seem at all obvious.

Hex

Piet Hein invented this game in 1942. It was reinvented by Nash
in 1948. People say that he had the idea while contemplating the
hexagonal tiling in the men's room of the Princeton mathematics
department. There were indeed hexagonal tiles there, but Nash
tells me that he doesn't recall finding them at all inspiring.

Hex is played between Black and White on a board of hexagons
arranged in a parallelogram, as in Figure 10. At the beginning of
the game, each player's territory consists of two opposite sides of
the board. The players take turns in moving, with White going
first. A move consists of placing one of your counters on a vacant
hexagon. The winner is the first to link their two sides of the board,
so Black was the winner in the game just concluded in Figure 10.

As in Chess, we can theoretically work out the players' maximin
payoffs using backward induction, but the method isn't practical

when the board is large. But we nevertheless know that White's maximin payoff is WIN. That is to say, the first player to move has a strategy that guarantees victory against any defence by the second player. How do we know this?

Note first that Hex can't end in a draw. To see this, think of the Black counters as water and the White counters as land. When all the hexagons are occupied, water will then either flow between the two lakes originally belonging to Black, or else the channel between them will be dammed. Black wins in the first case, and White in the second. So either Black or White has a winning strategy.

Nash invented a strategy-stealing argument to show that the winner must be White. The argument is by contradiction. If Black were to play a winning strategy, White could steal it using the following rules:

1. Put your first counter anywhere.

2. At later moves, first pretend that the last counter you played isn't on the board. Next pretend that all the remaining White counters are Black and all the Black counters are White.

3. Now make the move that Black would make in this position when using his winning strategy. If you already have a counter in this position, just move anywhere.

This strategy guarantees you a win, because you are simply doing what supposedly guarantees Black a win – but one move earlier. The presence on the board of an extra White counter may result in your winning sooner than Black would have done, but I guess you won't complain about that!

Since both players can't be winners, our assumption that Black has a winning strategy must be wrong. The winner is therefore White – although knowing this fact won't help her much when playing Hex on a large board, since finding White's winning strategy is an unsolved problem in the general case.

Notice that the strategy-stealing argument doesn't tell us anything at all about White's *actual* winning strategy. She certainly can't guarantee winning after putting her first counter just anywhere. If she puts her first counter in an acute corner of the board, you will probably be able to see why Black then has a winning strategy in the rest of the game.

It may also be fun to test your reasoning skills on a version of Hex with which Princeton mathematicians supposedly used to tease their visitors. An extra line of hexagons is added to the board so that White's two sides of the board become more distant than Black's. In the new game, not only is it Black who has a winning strategy, but we can write his winning strategy down. However, when visitors played as White against a computer, the board was shown in perspective on the screen to disguise its asymmetry. The visitors therefore thought they were playing regular Hex, but to their frustration and dismay, somehow the computer always won!

Deleting dominated strategies

Every time you throw out a bunch of choices at a node while carrying out a backward induction, you are discarding an equivalent bunch of pure strategies. From the point of view of the strategic form of the game you have reached at that stage, any strategy you discard is *dominated* by a strategy which is exactly the same except that it calls for a best choice to be made at the node in question.

If we exclude the case when two strategies always yield the same payoff, one strategy is dominated by another if it never yields a better payoff, no matter what strategies the other players may use. Thus *hawk* dominates *dove* in the Prisoner's Dilemma (but not in the Stag Hunt Game of Figure 18).

We can therefore mimic backward induction in a game by successively deleting dominated strategies in its strategic form. We can sometimes reduce a strategic form to just one outcome by this method even when not mimicking backward induction. The result will always be a maximin outcome in a two-person, zero-sum game. But what about games in general?

Any Nash equilibrium of a game you get by eliminating dominated strategies from a larger game must also be a Nash equilibrium of the larger game. The reason is that adding a dominated strategy to your options in a game can't make any of your current best replies into something worse. You may sometimes lose Nash equilibria as you delete dominated strategies (unless all the dominations are strict), but you can never eliminate *all* Nash equilibria of the original game.

Guessing games

If Alice trades on the stock market, she is hoping that the shares she buys will rise in value. Since their future value depends on what other people believe about them, investors like Alice are really investing on the basis of their beliefs about other people's beliefs. If Bob plans to exploit investors like Alice, he will need to take account of his beliefs about what she believes about what other people believe. If we want to exploit Bob, we will need to ask what we believe about what Bob believes about what Alice believes about what other people believe.

John Maynard Keynes famously used the beauty contests run by newspapers of his time to illustrate how these chains of beliefs about beliefs get longer and longer the more one thinks about the problem. The aim in these contests was to chose the girl chosen by most other people. Game theorists prefer a simpler Guessing Game in which the winners are the players who choose a number that is closest to two-thirds of the average of all the numbers chosen.

If the players are restricted to whole numbers between 1 and 10 inclusive, it is a dominated strategy to choose a number above 7, because the average can be at most 10, and $\frac{2}{3} \times 10 = 6\frac{2}{3}$. You therefore always improve your chances of winning by playing 7 instead of 8, 9, or 10. But if everybody knows that, nobody will ever play a dominated strategy, then we are in a game in which the players choose a number between 1 and 7 inclusive. The average in this game can be at most 7, and $\frac{2}{3} \times 7 = 4\frac{2}{3}$. So it is a dominated strategy to choose a number above 5.

It will be obvious where this argument is going. If it is common knowledge that no player will ever use a dominated strategy, then all the players must choose the number 1.

Common knowledge

Something is common knowledge if everybody knows it, everybody knows that everybody knows it, everybody knows that everybody knows that everybody knows it; and so on. If nothing is said to the contrary in a rational analysis of a game, it is always implicitly being assumed that both the game and the rationality of the players are common knowledge. Otherwise we wouldn't be entitled to use the idea of a Nash equilibrium to break into infinite regressions of the form: 'Alice thinks that Bob thinks that Alice thinks that Bob thinks ...'

I once watched a quiz show called *The Price is Right* in which three contestants guess the value of an antique. Whoever gets closest to the actual value is the winner. If the last contestant thinks the value is more than both the other two guesses, he should obviously raise the higher guess by no more than one dollar. Since this isn't what happens, we would be foolish to try to apply game theory to quiz shows on the assumption that it is common knowledge that the contestants are rational. It is therefore fortunate that the evolutionary interpretation of game theory doesn't require such strong assumptions.

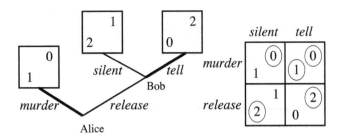

11. Kidnap

Subgame perfection

Daniel Ellsberg is best known for blowing the whistle on the Nixon administration's conduct of the war in Vietnam when he leaked the Pentagon Papers to the *New York Times* in 1971. In an earlier incarnation, he proposed the game of Kidnap.

Kidnap

Alice has kidnapped Bob. The ransom has been paid, and the question now is whether she should release him or murder him. Alice would prefer to release Bob if she could be sure that he wouldn't reveal her identity. Bob has promised to stay silent, but can she trust his promise?

Figure 11 shows a game tree for Kidnap together with a corresponding payoff table. Circling best replies reveals that there is only one Nash equilibrium, in which Alice murders Bob because she predicts that he will tell if released.

Deleting dominated strategies leads us to the same Nash equilibrium. Bob's strategy *tell* is always at least as good as *silent*. So we begin by deleting *silent*. In the game that remains, Alice's strategy *murder* is always at least as good as *release* (because Bob can only play *tell* in the reduced game). So we are left with only the Nash equilibrium (*murder*, *tell*).

Deleting dominated strategies in this way corresponds to using backward induction in the game tree. First thicken the branch in the game tree that represents Bob's best choice of *tell*. Now forget that Bob's inferior choice is there at all, and thicken the branch that represents Alice's best choice of *murder* in the game that remains. We can now see the equilibrium path that will be followed when Alice and Bob play the Nash equilibrium (*murder, tell*). In this case, a single thickened branch links the root of the tree to a leaf; in a bigger game, the equilibrium path will be a whole sequence of thickened branches that link the root to a leaf.

In games of perfect information like Kidnap, backward induction always leads to strategies that are not only a Nash equilibrium in the whole game, but also in all its subgames – whether they lie on the equilibrium path or not. Reinhard Selten shared a Nobel Prize with John Nash partly for introducing this class of equilibria. He first called them perfect, but later changed his mind about what perfection should mean. So now we call them *subgame perfect*.

Counterfactuals

Politicians like to pretend that hypothetical questions make no sense. As George Bush Senior said in 1992 when replying to a perfectly reasonable question about unemployment benefit: 'If a frog had wings, he wouldn't hit his tail on the ground.' But the game of Kidnap shows why hypothetical questions are the life blood of game theory – just as they ought to be the life blood of politics.

Rational players stick to their equilibrium strategies because of what they predict *would* happen if they *were* to deviate. The subjunctives in this sentence appear because we are talking about a counterfactual event – an event that isn't going to happen. Far from being irrelevant to anything real, such counterfactual events always arise when a rational decision is made. Why doesn't Alice ever step in front of a car when crossing the road? Because she predicts that if she did, she would be run over. Why does Alice murder Bob in Kidnap? Because she believes that he would tell on her if she didn't.

What would happen in subgames that won't be reached therefore matters. It is because of what would happen if they were reached that they aren't reached!

Changing the game?

Psychologists advise kidnap victims to try and build up a human relationship with their captors. If Bob could thereby persuade Alice that he cared sufficiently for her that his payoffs for remaining silent or telling were reversed, then we would be playing a different game that one might call Cosy Kidnap.

As Figure 12 shows, Cosy Kidnap has two Nash equilibria in pure strategies: (*murder*, *tell*) and (*release*, *silent*). The equilibrium (*murder*, *tell*) isn't subgame perfect any more, because it calls for Bob to make the inferior choice of *tell* in the subgame that is unreached in equilibrium because Alice actually chooses *murder*, but which would be reached if Alice were to choose *release* instead.

However, the new equilibrium (*release*, *silent*) is subgame perfect. It is therefore this equilibrium that will be played, provided that Alice is rational and knows that Bob is rational. If the payoffs are

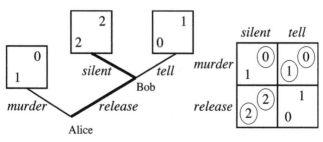

12. **Cosy Kidnap**

chosen according to the theory of revealed preference, then it is tautological that Bob would play *silent* rather than *tell* if Alice were to play *release*. Alice will therefore play *release* because she knows it will yield a higher payoff than *murder*.

The moral is that rationality sometimes tells us more than simply that Alice and Bob must play a Nash equilibrium.

Ultimatum Game

Reinhard Selten has a mischievous sense of humour, and it may be that he takes a delight in the controversy he created with his notion of a subgame-perfect equilibrium. He certainly added fuel to the fire when he proposed to his student Werner Güth that he run a laboratory experiment on the subject. The experiment was to see whether real people would play the subgame-perfect equilibrium in the Ultimatum Game. Selten predicted that they wouldn't – and he was right.

The Ultimatum Game is a primitive bargaining game in which a notional philanthropist has donated a sum of money for Alice and Bob to share if they can agree on how to divide it. The rules specify that Alice first makes a proposal to Bob on how to divide the money. He may accept or refuse. If he accepts, Alice's proposal is adopted. If he refuses, the game ends with both players getting nothing.

It is easy to apply backward induction to the game on the assumption that both players care only about getting as much money as possible. If Alice offers Bob a positive amount, he will say *yes*, because anything is better than nothing. The most that Alice will therefore offer is a penny. In a subgame-perfect equilibrium, Alice therefore scoops the pot.

However, laboratory experiments show that real people usually play fair. The most likely proposal is a fifty-fifty split. Proposals for an unfair split like seventy-thirty are refused more than half the time, even though the responder then gets nothing at all. This is the most replicated result in experimental economics. I have replicated it myself several times. It doesn't go away when the stakes are increased. It holds up even in countries where the dollar payoffs are a substantial fraction of the subjects' annual income. The result isn't entirely universal, but one has to follow anthropologists into remote parts of the world to find exceptions.

A new school of behavioural economists uses this result as a stick with which to beat their traditional rivals. They say that the data disprove the 'selfishness axiom' of orthodox economics. Their challenge is therefore to the hypothesis that people care only about money rather than to the logic of backward induction.

Actually, it isn't axiomatic in economics that people are relentlessly selfish. The orthodoxy is represented by the theory of revealed preference. Everybody agrees that money isn't everything. Even Milton Friedman used to be kind to animals and give money to charity. But it is also true that there are an enormous number of experiments showing that most subjects do eventually end up behaving as though they were primarily interested in maximizing their dollar payoffs in all but a few laboratory games. The Prisoner's Dilemma is the norm rather than an exception. So what is different about the Ultimatum Game?

I think that the answer lies in the fact that the rational and the evolutionary interpretations of an equilibrium diverge when applied to subgame-perfect equilibria.

The Ultimatum Minigame

In this simpified version of the Ultimatum Game, the philanthropist donates $4. Alice can make a fair or an unfair proposal to Bob. The fair offer is to split the money fifty-fifty. Bob automatically accepts the fair offer, but has the option of accepting

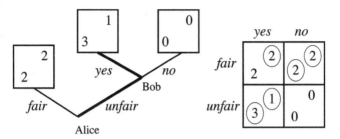

13. **Ultimatum Minigame. Apart from the labels of the available actions and some inconsequential changes in the payoffs, the game is the same as Cosy Kidnap**

or refusing the unfair offer, which assigns \$3 to Alice and only \$1 to Bob. Figure 13 shows the game tree and payoff table for the Ultimatum Minigame. Its analysis is the same as in Cosy Kidnap, although here the logic of the argument is controversial because critics don't like where it leads.

The subgame-perfect equilibrium is (*unfair, yes*). Like Cosy Kidnap, the game also has another Nash equilibrium: (*fair, no*). In fact, it has lots of Nash equilibria in which Alice chooses *fair* because Bob is planning to use a mixed strategy in which he says *no* to the unfair offer with a sufficiently high probability.

The reason that we need to worry about Nash equilibria that aren't subgame perfect is that we haven't any reason to suppose that an evolutionary process will necessarily converge on the subgame-perfect equilibrium. If the subjects are learning by trial and error which equilibrium to play, they might therefore learn to play any of the Nash equilibria of the Ultimatum Minigame.

Figure 14 shows two different evolutionary processes in the Ultimatum Minigame. One is the best-reply dynamics we encountered earlier; the other is the more complicated *replicator dynamics*, which is usually regarded as a superior toy model of an adjustment process (see Evolutionary Stability, Chapter 8).

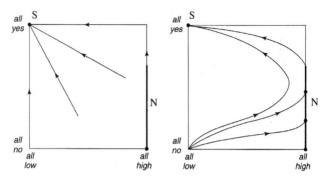

14. Evolutionary adjustment in the Ultimatum Minigame. The subgame-perfect equilibrium is *S*. The other Nash equilibria lie in the set *N*. The latter all require the use of the weakly dominated strategy *no*, but *N* still has a large basin of attraction in the case of the replicator dynamics

The best-reply dynamics converge on the subgame-perfect equilibrium, but this isn't necessarily true of the replicator dynamics. The set of Nash equilibria in which Alice plays *fair* has a large basin of attraction in Figure 14.

Evolution doesn't care that Bob's choice of *no* is weakly dominated in all of these equilibria. It is true that *yes* is always better than *no* provided that Alice sometimes plays *unfair*, but the evolutionary pressure against *unfair* can be so strong that it disappears altogether. Once it has gone, *no* can survive, because Bob is then indifferent between *yes* and *no*.

Fair conventions

We now have an explanation of the experimental data in the Ultimatum Game that doesn't require assigning different preferences to the subjects than they reveal when playing the Prisoner's Dilemma in the laboratory.

In real life, Bob would be stupid to knuckle under when made an unfair offer, because he can't afford to acquire a reputation for being a soft touch. We therefore operate a convention in which Alice is often refused if she makes an unfair offer. Subjects bring this convention into the laboratory without realizing either that it coordinates behaviour on an equilibrium in the game of life, or that the game they are asked to play in the laboratory is very different from the real-life games for which the convention is adapted.

When subjects start by playing fair in the Prisoner's Dilemma, evolutionary pressures immediately start modifying their behaviour, because the only Nash equilibrium in the Prisoner's Dilemma precludes any cooperation. The Ultimatum Game differs from the Prisoner's Dilemma in having many Nash equilibria. Any split whatever of the available money corresponds to a Nash equilibrium, for the same reason that the same is true in the Ultimatum Minigame. When Alice and Bob begin by playing fair in the Ultimatum Game, there are no obvious evolutionary pressures urging them towards the subgame-perfect equilibrium. We therefore don't need to invent some reason why they don't move much from where they started.

Game theorists are happy for behavioural economists to make the case against selfishness. How else are we to explain why Milton Friedman contributed to charity? But they make two errors when they say: 'Game theory predicts the subgame-perfect equilibrium in the Ultimatum Game.' The first is that game theory assumes that players necessarily maximize money. The second is that rational and evolutionary game theory always predict the same thing.

Refinements

Evolution doesn't always select subgame-perfect equilibria, but it remains rational for Alice to solve the Ultimatum Minigame by backward induction when the payoffs are determined by the theory of revealed preference. The standard assumption that Alice knows that Bob is rational is essential for this purpose, because Alice needs to be sure that Bob's behaviour will be consistent with the payoffs assigned to him.

Does our standard assumption that the rationality of the players is common knowledge imply that a subgame-perfect equilibrium path will be followed in *any* finite game of perfect information? Bob Aumann says *yes,* and one might think that he should know, since he won his Nobel Prize partly for making common knowledge into an operational tool. But examples like Selten's Chain Store paradox continue to keep the question open.

Chain Store paradox

The Ultimatum Minigame can be reinterpreted as a game in which Alice is threatening to open a store in a town where Bob already runs a similar store. We just need to relabel Alice's strategies as *out* and *in,* and Bob's as *acquiesce* and *fight*. Fighting consists of initiating a price war, which is bad for both players. Selten's paradox arises when Bob runs a chain of stores in a hundred towns and Alice is replaced by a hundred possible rivals threatening to set up a rival store in each town.

Just as in the Ultimatum Minigame, backward induction in the 100th game says that the 100th rival will enter the market, and Bob will acquiesce. What happens in the 100th game is therefore determined independently of what happens in previous games, and so exactly the same argument applies in the 99th game. Continuing in this way, we end up with the conclusion that the rival will always enter and Bob will always acquiesce. But wouldn't Bob do better to fight the first few entrants so as to discourage entry in the remaining towns?

The game tree of Figure 15 is a simplification in which there are only two towns and the rival is always Alice. If she enters the first town, Bob can acquiesce or fight. If she later enters the second

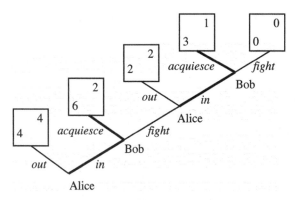

15. A simplified Chain Store paradox. Apart from the labels of the available actions, the subgame rooted at Alice's second move is identical to the Ultimatum Minigame

town, he can again acquiesce or fight. If Alice stays out of the first town, we simplify by assuming that she necessarily stays out of the second town. Similarly, if Bob acquiesces in the first town, Alice necessarily enters the second town, and Bob again acquiesces.

The thickened lines in Figure 15 show the result of applying backward induction. If the great book of game theory recommended following the subgame-perfect equilibrium path, it would therefore be right for Alice to enter both towns and for Bob to acquiesce each time. But will Alice and Bob follow the book's advice? To explore this question, put yourself in Bob's position at his first move.

Alice has just entered the first town as recommended by the book, but what would she do if her second move were reached? The answer depends on what she predicts Bob would do if his second move were reached. If Alice knew that Bob were rational, then she would predict that he would acquiesce. She should then enter, and so Bob should acquiesce at his first move, as required by backward induction. But Alice *wouldn't* know that Bob is rational at her second move, because a rational Bob wouldn't have fought at his first move if the great book of game theory were right about what is rational!

Alice began the game believing Bob to be rational, but if he plays in a manner that is inconsistent with his preferences by fighting in the first town, her belief will be refuted. And who knows what she might believe after such a counterfactual event? Selten's original version of the paradox has 100 stores, because the common-sense answer after Bob has fought in 50 towns is that he is likely to fight in the 51st as well. But then the backward induction argument collapses.

The paradox doesn't cast doubt on backward induction as a way of finding the maximin payoffs in two-person, zero-sum games. Nor does it create a problem for the rationality of backward induction in games like Kidnap or the Ultimatum Game. The players' initial belief that everyone is rational would still be refuted if someone were to diverge from the equilibrium path, but this fact causes no problem in these short games. But how are we to respond to the paradox in longer games?

Typos

Subgame-perfect equilibria are said to be a *refinement* of the Nash equilibrium concept. They are safe to use whenever the circumstances make it sensible for the players to continue behaving as though it is common knowledge that they are all rational even though one or more irrational moves have been made. A whole bestiary of even more refined refinements has been created for use in games of imperfect information. These are based on various different ideas about what beliefs would make sense in the counterfactual event that a rational player were to play irrationally. If George Bush Senior were to read the literature, it would make his head swim! Fortunately, this phase in the history of game theory is effectively over – although applied economists continue to appeal to whichever refinement in the bestiary comes closest to confirming their own prejudices.

My own take on these issues is that we should follow Reinhard Selten's common-sense approach, which eliminates the need to interpret counterfactuals at all. He recommends that we build enough chance moves into the rules of our games to remove the possibility that players will find themselves trying to explain the inexplicable. In the simplest such models, the players are assumed to make occasional mistakes. Their hands tremble as they reach for the rational button and they press an irrational button by mistake. If these mistakes are independent transient errors – like typos – that have no implications for mistakes that might be made in the future, then the Nash equilibria of the game with mistakes converge on subgame-perfect equilibria of the game without mistakes as we allow the frequency of mistakes to get very small.

Selten tried to downgrade subgame-perfect equilibria because he decided that the limits of Nash equilibria in trembling-hand games are what really deserve to be called perfect. But the rest of the world only concedes that such equilibria are trembling-hand perfect.

Thinkos

The reason that other game theorists were unwilling to endorse Selten's new definition can perhaps be traced to doubts about the generality of his trembling-hand story. If we want a rational analysis of a game to be relevant to the behaviour of real people trying to cope intelligently with complex problems, we have to face up to the fact that their mistakes are much more likely to be 'thinkos' than 'typos'.

For example, nobody would think it reasonable to explain why the owner of a chain of stores initiated a price war in 50 successive towns by saying that he always meant to instruct his managers to acquiesce in the entry of a rival, but somehow always sent the wrong message by mistake. The only plausible explanation is that he has a policy of fighting entry, and hence is likely to fight in the 51st town whether this is foolish or not.

When chance moves are introduced that allow for such thinkos to occur, the Nash equilibria of the game with mistakes needn't converge on a subgame-perfect equilibrium of the game without mistakes. So Nash equilibria of the game without mistakes can't routinely be thrown away as irrelevant to a rational analysis. But nor do we want to scrap backward induction. All Nash equilibria of the game with mistakes are automatically subgame-perfect because the mistakes ensure that every subgame is always reached with positive probability. Backward induction is therefore a useful tool when locating these equilibria.

A moral?

The lesson I draw from the refinement controversy is that game theorists went astray by forgetting that their discipline has no substantive content. Just as it isn't our business to say what people ought to like, so it isn't our business to say what they ought to believe. We can only say that if they believe this, then they would be inconsistent not to believe that. If we can't analyse a game on such consistency principles alone, then more information about the players and their environment needs to be added to the game until we can.

Chapter 4
Conventions

There is no problem about which Nash equilibrium should be regarded as the rational solution of a two-person, zero-sum game, because any pair of maximin strategies is always a Nash equilibrium in which the players get their maximin payoffs. But things can be very different in games that aren't zero sum.

For example, in the Battle of the Sexes, the maximin payoff for both players is two-thirds. This happens to be the same as the payoffs they both get in the game's mixed equilibrium, but their maximin strategies aren't equilibrium strategies. Moreover, Alice and Bob's payoffs in both pure equilibria of the game are much bigger than their maximin payoffs. So what should they do?

The Driving Game makes it obvious that there isn't any point in looking for a strictly rational answer. Any argument that might be offered in favour of everyone driving on the left would be an equally good argument for everyone driving on the right. People sometimes say that the rational solution must therefore be the mixed equilibrium in which everyone decides whether to drive on the left or right at random, but this proposal seldom garners much support!

To solve the Driving Game, we need a commonly accepted *convention* as to whether we should drive on the left or the right.

The fact that such a convention may be entirely arbitrary is reflected in the fact that some countries have adopted the convention of driving on the left and others of driving on the right.

Focal points

Societies sometimes choose conventions deliberately, as when Sweden switched from driving on the left to driving on the right in the early hours of 1 September 1967. However, one should perhaps think of Sweden's problem on this occasion as a multi-player version of the Battle of the Sexes, with some players preferring the traditional equilibrium and others preferring the equilibrium used by the rest of continental Europe. Rationality alone can't settle such differences over how to solve equilibrium selection problems, but the convention in Sweden is to follow the guidance of its democratically elected government. On the other hand, it is only necessary to observe the chaos at traffic signals in Naples to realize that the guidance of a democratically elected government isn't enough to guarantee that a convention will be honoured.

Tom Schelling

What happens when no obvious convention is in place? Tom Schelling ran a number of experiments in the 1950s which show that we aren't so helpless as you might think at first sight. He says that the conventions that people invent when asked questions like the following are focal points. Most people are surprised both at their success in locating focal points, and at the arbitrary nature of the contextual cues to which they find themselves appealing. An important lesson is that the context in which games appear – the way a game is *framed* – can make a big difference to how real people play them.

1. Two players independently name *heads* or *tails*. They win nothing unless both say the same, in which case each wins $100. What would you say?

2. You are to meet someone in New York tomorrow, but no arrangements have been made about where or when the meeting is to take place. Where will you go? At what time?

3. Alice, Bob, and Carol must each independently write down the letters *A*, *B*, and *C* in some order. They all get nothing unless they choose the same order, in which case the player whose initial is first gets $300, the player whose initial is second gets $200, and the player whose initial is third gets $100. What would you do if you were Carol?

4. Alice and Bob are each given one of two cards. One card is blank and the other is marked with a cross. A player can mark a cross on the first card or erase the cross on the second. Nobody wins anything unless there is one and only one cross on the two cards when they are handed in. The player who hands in the card with the cross then wins $200 and the player who hands in the blank card wins $100. What would you do if given the blank card?

5. A philanthropist donates $100 to Alice and Bob – provided they can agree on how to divide it. Each player is independently required to claim a share. If the shares sum to more than $100, nobody gets anything. Otherwise each player receives the amount that he or she claimed. How much would you claim?

6. Alice loses $100 and Bob finds it. Bob is too honest to spend the money, but is unwilling to return it unless suitably rewarded. What reward would you offer to Bob if you were Alice? What reward would you offer if Bob had already refused $20? What reward would you offer if Alice and Bob had watched a television programme together the previous evening on which some guru announced that the fair split in such circumstances is for Bob to get a reward of one-third of the total amount?

Most people say *heads* in the first question, because it is conventional to say *heads* before *tails* when both are mentioned. How well people do in the second question depends on their familiarity with New York. Schelling asked New Englanders, who strongly favoured Grand Central Station at noon. In the third question, Carol usually recognizes that alphabetical order is so focal that she has to say *ABC*, although she will then get the lowest payoff of the three players. In the fourth question, the *status quo* is focal, and most people therefore choose to do nothing. In the fifth question, a fifty-fifty split is almost universal. The sixth question is more challenging. People usually manage to coordinate effectively only after hearing about the guru, in which case they nearly always take his advice.

How much is conventional?

Daily life largely consists of playing a multitude of coordination games with those around us. When young people learn how to play these coordination games by emulating the successful players in their environment, they usually don't notice that they are playing a game at all. They learn whatever convention is current in their society without appreciating that the convention wouldn't survive in the long run unless it coordinated behaviour on an equilibrium. When the convention itself evolved so long ago that its origins are lost in the mists of time, it can even become conventional to deny that the convention is conventional. It then becomes impossible to recognize that other societies may be playing essentially the same game as us, but that their different social history has led to a different equilibrium of the game becoming focal.

David Hume was the first philosopher to say out loud that many of our rules of social behaviour are no more solidly founded than the convention we use to select an equilibrium in the Driving Game. In his *Treatise* of 1739, he famously says:

> Two men who pull the oars of a boat, do it by an agreement or convention, although they have never given promises to each other. Nor is the rule concerning the stability of possessions the less derived from human conventions, that it arises gradually, and acquires force by a slow progression ... In like manner are languages gradually established by human conventions without any

16. David Hume

promise. In like manner do gold and silver become the common measures of exchange, and are esteemed sufficient payment for what is of a hundred times their value.

Most people have no difficulty in accepting the conventional nature of language or money, but draw the line when philosophers like Hume suggest that the same is true in sensitive subjects like ethics or religion. Sometimes their opposition to moral relativism or evolutionary biology is so strong that they feel the need to throw out game theory as well. But, whatever you may think of Hume's bathwater, game theory is a baby worth clasping to your bosom.

Game theory can never be a threat to any consistent religious or ethical system, because it has no more substantive content than arithmetic or logic. It only says that some propositions aren't consistent with other propositions. Like arithmetic or logic, it can therefore be used on either side of any argument.

Many game theorists are deeply religious – notably Bob Aumann, who shared a Nobel Prize with Tom Schelling in 2006. Steve Brams has even written a book using game theory to make theological points. I am a sceptic myself, but my *Natural Justice* accepts that some fairness principles are universal in the human species. In brief, the only folk who need fear the use of game theory are those whose beliefs are inconsistent.

Bad conventions

The mixed equilibrium in the Driving Game isn't at all efficient, since players who use it will end up in a stand-off half the time. But it is an equilibrium nevertheless, and hence available as a possible convention. I used to say that it is a convention that has never actually emerged anywhere in the world, until corrected by some Turks, who observed that I had obviously never visited Turkey. But I have now, and I see what they mean.

Schelling's Solitaire is a toy model intended to show how cultural evolution can easily bring about such socially undesirable conventions without any need for an evil genius plotting the downfall of society in the background. It is played on a Chessboard with Black and White counters. Each counter represents a

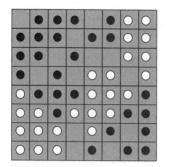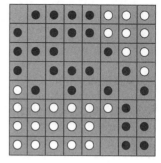

17. Schelling's Solitaire

householder. The square the counter occupies represents his or
her house. The surrounding squares (up to eight) represent the
householder's neighbourhood. So a counter on one of these
squares is a neighbour.

Each counter is sensitive to the colours of its neighbours. Whites
wish one half or more of their neighbours to be White. Blacks wish
one third or more of their neighbours to be Black. You operate the
evolutionary process by moving discontented counters to squares
on which they are content until nobody who wants to move has
anywhere they want to go. Schelling recommends starting with
Black counters on all the Black squares of the Chessboard and
White counters on all the White squares. You remove some of
these at random and then start the process. In Figure 17,
12 counters were removed.

The two configurations shown in Figure 17 are equilibrium
outcomes of the process. They differ because there is some
randomness in the initial configuration and in your choice of
which discontented counter to move next. But the equilibrium
that emerges nearly always has Black and White counters
occupying segregated neighbourhoods. It is worth playing
Schelling's Solitaire a few times to get a feeling for how inexorable
the separation process can be. Everyone in the model would be

content to live in a mixed neighbourhood but they end up with a convention in which they are segregated.

Social dilemmas

Schelling's Solitaire shows how easy it can be for a convention to get established that nobody likes. Following the sociologist Vilfredo Pareto, economists say that outcomes are inefficient when there are other outcomes that everybody likes better. But if we realize that we are operating an inefficient convention, aren't we being collectively irrational not to move to an efficient one?

Perhaps the most important role for game theory is to insist that any reform needs to coordinate behaviour on an *equilibrium* if it is to survive in the long run. If there is no satisfactory equilibrium to which we can move, as in Schelling's Solitaire, inventing a new kind of rationality that somehow whitewashes over the inconsistencies in individual behaviour implied by out-of-equilibrium play can only make things worse. One only has to look at the long history of failed utopias to see why.

Karl Marx is a major culprit. In treating Capital and Labor as monolithic players in a mighty game, he failed to see that the cohesion of a coalition depends on the extent to which it succeeds in satisfying the aspirations of its individual members. The same is true when a whole society is treated as though it were a single individual written large. This isn't to deny that group solidarity can sometimes temporarily triumph over individual incentives – even when there is no prospect of blacklegs being punished by the comrades they betray. Nor is to deny that we would arguably all be better off if we identified more often with some conception of the common good. Such behaviour is certainly selfless or even saintly, but we defraud ourselves if we insist that more selfish behaviour is somehow irrational.

As the grumpy philosopher Thomas Hobbes explained long ago:

> Bees or Ants, live sociably one with another … and therefore some
> man may perhaps desire to know why Mankind cannot do the same.
> To which I answer … amongst these creatures, the common good
> differeth not from the private.

In game theory terms, Hobbes is saying that the games that social
insects play with each other are games of pure coordination, but I
guess that most people would agree with me that the same is
seldom true of human beings.

The errors I am pointing out here are typical of intellectuals of the
left, but intellectuals of the right need not congratulate
themselves. They typically make the complementary error of
overlooking the possibility that there may be more efficient
equilibria than the equilibrium we are currently operating.

What game theory can contribute to such debates is a framework
within which one can realistically discuss what is or is not possible
for a society. What equilibria are available in the game we are
playing? Is there an equilibrium we all like better than the
equilibrium we are currently playing? If we don't like any of the
available equilibria, can we change the rules of the game or the
players' preferences somehow?

Suppose everybody behaved like that?

Social psychologists say that a situation in which achieving an
efficient outcome conflicts with the incentives of the individual
members of a group is a social dilemma. The Prisoner's Dilemma
is the archetypal example.

You can usually tell that you are in a social dilemma by the fact
that your mother would register her disapproval of any hawkish
inclination on your part by saying, 'Suppose everybody behaved

like that?' Immanuel Kant is sometimes said to be the greatest philosopher of all time, but he too thought that it couldn't be rational to do something if it would be bad if everybody did it. As his famous categorical imperative says: 'Act only on the maxim that you would will to be a universal law.'

For example, when waiting at an airport carousel for our bags, we would all be better off if we all stood well back so that we could see our bags coming. But if everybody else does that, it profits each individual to edge forward a little, and so we all end up straining our necks to peer over a wall of backs.

We would similarly all benefit if we turned our air conditioners down when a brown-out is threatened, or if we didn't use our lawn sprinklers in a drought. The same applies when people stand up at a football match, or when they conduct their business in slow motion after reaching the head of a long line.

When large numbers of anonymous folk play such social dilemmas, Kant and your mother are right to predict that things will work out badly if everybody responds to their individual incentives. But urging people to behave better in such situations is seldom effective. Why should you lose out by paying heed to your mother when everybody else is ignoring theirs?

Tragedy of the Commons

The everyday social dilemmas described above are irritating, but some social dilemmas spell life or death for those who must play them. A toy example is called the Tragedy of the Commons by political scientists.

A hundred families keep goats that graze on some common land. Total milk production is maximized with a thousand goats in all.

How many goats should each family keep to maximize its own milk production?

At first sight, the answer seems to be ten, but it isn't an equilibrium for each family to keep ten goats. If all the other families keep ten goats, your family's optimal strategy isn't to do the same. You will do better by grazing one goat more, because your family will enjoy all the benefit from the extra goat, while its cost in terms of less grass for the other goats to eat will be shared by the whole community. Families will therefore add extra goats to their herd until the common is reduced to a desert. But this outcome is very inefficient indeed.

The Tragedy of the Commons captures the logic of a whole spectrum of environmental disasters that we have brought upon ourselves. The Sahara Desert is relentlessly expanding southward, partly because the pastoral peoples who live on its borders persistently overgraze its marginal grasslands. We pump carbon dioxide into the atmosphere as though there were no tomorrow. We poison our rivers. We jam our roads with cars. We fell the rainforests. We have plundered our fisheries until some fish stocks have reached a level from which they may never recover.

Game theorists get a lot of stick for denying that the individual behaviour that leads to such disasters is irrational. Our critics ask how it can possibly be rational for a society to engineer its own ruin. Can't we see that everybody would be better off if everybody were to grab less of the common resource? The error in such reasoning is elementary. A player in the human game of life isn't some abstract entity called 'everybody'. We are all separate individuals, each with our own aims and purposes. Even when our capacity for love moves us to make sacrifices for others, we each do so in our own way and for our own reasons. If we pretend otherwise, we have no hope of ever getting to grips with the Tragedy of the Commons.

Stag Hunt Game

Jean-Jacques Rousseau, the prophet of the French Revolution, accepted that political games are seldom games of pure coordination. His solution was to convert them into games of pure coordination by changing our preferences: 'If you would have the general will accomplished, bring all the particular wills into conformity with it.'

Game theorists overlook the impracticality of this radical programme and focus instead on his parable of a stag hunt. Alice and Bob agree to cooperate in hunting a stag, but when they separate to put their plan into action, each may be tempted to abandon the joint enterprise by the prospect of bagging a hare for themselves.

Only slight changes in the payoffs of the give-and-take version of the Prisoner's Dilemma of Figure 7 are necessary to obtain the Stag Hunt Game of Figure 18, but they are enough to make it a Nash equilibrium for both players to play *dove*.

The Stag Hunt Game therefore looks like a game that doesn't create a social dilemma. If we find ourselves operating the inefficient Nash equilibrium in which Alice and Bob both play *hawk*, we can shift to the efficient Nash equilibrium in which they both play *dove*. However, the payoffs in the Stag Hunt Game have been carefully chosen to make such a shift hard to manage.

The basin of attraction of the inefficient equilibrium is large and that of the efficient equilibrium is small. So it is difficult for evolution to get us out of the basin of attraction of the inefficient equilibrium and into the basin of attraction of the efficient equilibrium. It is true that we aren't animals who have to wait for the slow forces of evolution to establish a new convention. We can talk to each other and agree to alter the way we do things. But can we trust each other to keep any agreement we might make?

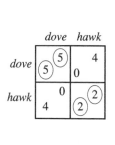

	dove	hawk
dove	(5) / 5	4 / 0
hawk	0 / 4	(2) / 2

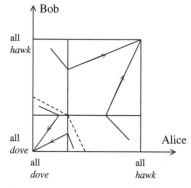

18. The Stag Hunt Game. The diagram on the right shows that the basin of attraction for the Nash equilibrium (*dove, dove*) is much smaller than the basin of attraction for the Nash equilibrium (*hawk, hawk*). The basin of attraction for the mixed Nash equilibrium in which *dove* is played two-thirds of the time is just the broken line

Variants of the Stag Hunt Game are used by experts in international relations under the name of the Security Dilemma or the Assurance Game to draw attention to the problems that can arise even when the players are rational.

Suppose that the current convention is to play *hawk*, but Alice seeks to persuade Bob that she plans to play *dove* in the future, and so he should follow suit. Will he be convinced? Game theorists think not. The reason is that whatever Alice is actually planning to play, it is in her interests to persuade Bob to play *dove*. If she succeeds, she will get 5 rather than 0 if she is planning to play *dove*, and 4 rather than 2 if she is planning to play *hawk*. Rationality alone therefore doesn't allow Bob to deduce anything about her plan of action from what she says, because she is going to say the same thing no matter what her real plan may be! Alice may actually think that Bob is unlikely to be persuaded to switch from *hawk* and hence be planning to play *hawk* herself, yet still try to persuade him to play *dove*.

Trust?

This Machiavellian story shows that attributing rationality to the players isn't enough to resolve the equilibrium selection problem – even in a seemingly transparent case like the Stag Hunt Game. The standard response is to ask why game theorists insist that it is irrational for people to trust one another. Wouldn't Alice and Bob both be better off if both had more faith in each other's honesty?

Nobody denies that Alice and Bob would be better off if they trusted each other. Nor do game theorists say that trust is irrational. They only say that it isn't rational to trust people without a good reason: that trust can't be taken on trust. For example, no Neapolitan is likely to trust his fellow drivers to start honouring traffic signals just because some authority figure says that they should.

So how can we ever move from one equilibrium to another? The collapse of the Soviet Empire provides a magnificent case study. Some Eastern European countries successfully followed the example of Sweden in the Driving Game by switching from a command economy to a market economy more or less overnight. They thereby minimized the opportunity for matters to slide out of control – as in Gorbachev's Russia – while the system was out of equilibrium during the interregnum.

But it would be just as much a mistake to deduce from the Stag Hunt Game that gradual transitions between conventions are never possible as it would be to deduce from the Prisoner's Dilemma that the same is true of rational cooperation. Neither game is adequate as a model of how whole societies work. They are just toys, invented to make a particular point.

Chapter 5
Reciprocity

If we want to understand a whole society, we can't neglect the role of reciprocity, which philosophers from Confucius to Hume have identified as the mainspring of human sociality. If they are right, then we all play our part in maintaining a complex network of reciprocal arrangements with those around us. But we understand how the system works no better than the physics we use when riding a bicycle.

Game theory offers some insight into the nuts and bolts of such self-policing understandings. How do they work? Why do they survive? How much cooperation can they support?

Repeated games

In a one-shot game, Alice can't promise to scratch Bob's back tomorrow if he will scratch her back today, because we have implicitly assumed that they are never going to meet again. The simplest setting in which reciprocity can arise requires the same players to play the same game over and over again.

Repetition with a fixed horizon

The Prisoner's Dilemma epitomizes the fact that cooperation needn't be rational. Does this unpleasant conclusion go away if

Alice and Bob play repeatedly? If it is common knowledge that Alice and Bob are to play the Prisoner's Dilemma every day for the next week, backward induction says that the answer is *no*. Politicians get dismissed as lame ducks near the end of their period in office for much the same reason.

On Saturday – the last day of the week – Alice and Bob will be playing the regular Prisoner's Dilemma, in which it is rational to play *hawk*. On Friday, they will therefore know that nothing they do today can affect what will happen tomorrow. So they will play *hawk* on Friday. Working backward through each day of the week, we find that rational players will always play *hawk*. (There are also Nash equilibria that aren't subgame perfect, but all these also require that *hawk* is played on the equilibrium path.)

Indefinite repetition

Should we conclude that rational cooperation is impossible even when the Prisoner's Dilemma is repeated? This would be a rash conclusion, because it isn't realistic to assume that Alice and Bob are sure that they will never interact again after next Saturday. In real life, relationships are nearly always open-ended. So what happens in the repeated Prisoner's Dilemma if it isn't common knowledge that Alice and Bob will never meet again? The answer is that rational cooperation now becomes feasible.

In the simplest toy model, Alice and Bob always believe that there is some positive probability that they will play the Prisoner's Dilemma at least one more time, no matter how many times they may have played in the past. If this probability is large enough and the players care enough about their future payoffs, then the repeated game has many Nash equilibria. In some of these, *dove* is always played on the equilibrium path.

To see this, it is enough to look at the GRIM strategy, which requires Alice always to play *dove* in the indefinitely repeated

Prisoner's Dilemma unless Bob ever plays *hawk*. If he does play *hawk*, the GRIM strategy says that Alice should retaliate by permanently switching to *hawk* herself. If both players use the GRIM strategy, neither will provoke the other into playing *hawk*, and so they will both play *dove* all the time. But is the pair (GRIM, GRIM) a Nash equilibrium?

All best replies to the GRIM strategy tell Bob never to be the first to play *hawk*. If he does play *hawk*, the best payoff stream he could get in the future is $3, 1, 1, 1, \ldots$, which is worse than the payoff stream of $2, 2, 2, \ldots$ he gets by always playing *dove*. As the GRIM strategy always calls for the play of *dove* when matched against itself, it follows that the choice of GRIM by Bob is a best reply to the choice of GRIM by Alice. Since the same is true of Alice, the pair (GRIM, GRIM) is a Nash equilibrium for the indefinitely repeated Prisoner's Dilemma.

Punishment

Critics who mistakenly believe that game theory denies that people are naturally altruistic sometimes take umbrage at the idea that cooperation supposedly can't work without the threat of punishment. They particularly dislike the GRIM strategy because it punishes any deviation from the equilibrium path with relentless determination.

Such critics are right to the extent that the threat of punishment is intrinsic to *reciprocal* cooperation. If Alice tells Bob that she will scratch his back if he will scratch hers, the implication is that she won't scratch his back if he won't scratch hers. People don't usually provide a service unless they expect to get something in return. If the service isn't reciprocated, then it will be withdrawn. Sometimes, a disservice will be offered instead. However, extravagant punishments like those built into the GRIM strategy are only encountered in extreme circumstances in real life. Everyday punishments are more usually proportional to the offence.

19. Reciprocal grooming by chimps

We are so habituated to responding appropriately to the small
punishments that are provoked by our small offences that we
seldom even notice that we are responding at all. Subliminal
signals from those around us are automatically translated into
behaviour without any conscious control. No stick is commonly
flourished. What happens most of the time is that the carrot is
withdrawn a little. Shoulders are turned slightly away. Greetings
are imperceptibly gruffer. Eyes wonder elsewhere. These are all
warnings that your body ignores at its peril, because they signal
that more serious social exclusion will follow if you don't mend
your ways.

Altruism?

The fact that game theorists think that more cooperation is
reciprocal than is generally appreciated doesn't imply that they
hold that cooperation is impossible without reciprocity. If people
have sufficiently altruistic preferences, then rational cooperation
ceases to be problematic even in one-shot games. For example, if
Alice and Bob have utilitarian preferences that make them want to
maximize the sum of both their payoffs rather than their own
individual payoffs, then it would be a Nash equilibrium for both to

play *dove* in the Prisoner's Dilemma. (We shall encounter exactly this case when Alice and Bob are identical twins in the Hawk-Dove Game in Chapter 8.)

How much people care about each other is an empirical question on which game theory is necessarily silent. My own view is that although the human cocktail obviously contains more than a smidgen of Dr Jekyll, I wouldn't join a utopia that denies the existence of Mr Hyde.

Such utopias sometimes work well enough to begin with, but the original sweetness and light notoriously erode as the players unconsciously respond to their incentives. Here, for example, is an IRS Commissioner explaining why a survey showed that the percentage of the public who thought it OK to cheat on their taxes was up from 11% to 17% over the previous five years: 'It's a basic sense of fairness. Somebody out there is complying with the law, and they see others doing things, and over time, they feel like chumps' (Mark Everett in *USA Today,* 8 April 2004). So the IRS continues to audit on the assumption that nearly everybody will eventually find an excuse to cheat if they don't provide adequate disincentives.

Folk theorem

Can strategies other than GRIM support rational cooperation in the indefinitely repeated Prisoner's Dilemma? What of rational cooperation in other repeated games?

Although game theory's answer to these questions is called the folk theorem, there is no Professor Folk. After Nash published his ideas on Nash equilibrium, Bob Aumann found that everybody in the business already seemed to know about the implications for repeated games, and so he decided that his thoughts on the subject should be regarded as folk wisdom.

David Hume had already explained how reciprocity works in 1739, but I don't suppose that Aumann knew anything about his work. The biologist Robert Trivers was equally ignorant of Aumann's ideas when he reinvented them under the heading of reciprocal altruism 20 years later. It was only with the publication of Axelrod's *Evolution of Cooperation* in 1984 that the idea finally stopped being rediscovered – in much the same way that America ceased being discovered after the voyage of Columbus in 1492.

The Trust Minigame

As a small child, I remember wondering why shopkeepers hand over the goods after being paid. Why don't they just pocket the money?

Economists call this the hold-up problem. My favourite example is the Antwerp diamond market. Traders hand over enormously valuable diamonds for inspection without even asking for a receipt. Why don't they get cheated? I found the neatest explanation in the *New York Times* of 29 August 1991. When asked why he could rely on the honesty of the owner of the antique store that sold his finds on commission, a dealer unfamiliar with the GRIM strategy replied: 'Sure I trust him. You know the ones to trust in this business. The ones who betray you, bye-bye.'

The Trust Minigame is a toy game that highlights these questions of trust and reputation. When Alice delivers a service to Bob, trusting him to reciprocate by making a payment in return, their predicament is essentially the same as in the game Kidnap of Figure 11. To see why, just relabel Alice's *release* strategy as *deliver* and Bob's *silent* strategy as *pay*.

Since Kidnap has a unique subgame-perfect equilibrium, the same is true of the Trust Minigame. Alice doesn't deliver the service, because she predicts that Bob won't pay. But the folk theorem tells us that all payoff pairs in the heavily shaded region of Figure 20

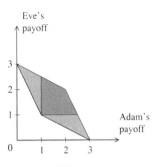

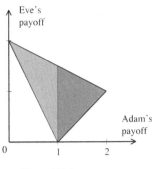

Prisoner's Dilemma · Trust Minigame

20. The folk theorem. The shaded regions show the cooperative payoff regions for the Prisoner's Dilemma and the Trust Minigame. These are the sets of payoff pairs on which the players could agree if pre-play bargains were enforceable. The deeply shaded triangles show the payoff pairs per game that the folk theorem shows to be available as equilibrium outcomes in the repeated versions of the two games when the players are sufficiently patient

are Nash equilibrium outcomes of the indefinitely *repeated* game – including the payoff pair (2, 1) that arises when Alice always delivers and Bob always pays.

To understand why the folk theorem works, it is useful to ask what possibilities would be open to Alice and Bob if they were to bargain in advance about how to play the one-shot Trust Minigame. One possibility is that they might agree on any of the three payoff pairs in the table of Figure 11. These payoff pairs are located at the corners of the shaded triangle in Figure 20. The remaining points in the triangle can be achieved as compromises obtained by tossing coins or taking turns. For example, the payoff pair that results if Alice and Bob agree that she will always deliver but he will only pay half the time lies midway between the pairs (0, 2) and (2, 1). Since it consists of all possible compromises on which Alice and Bob might agree, the shaded triangle is called the cooperative payoff region for the Trust Minigame.

The problem with the bargaining story is that it won't work without some kind of external agency willing and able to enforce any contracts that Alice and Bob may write. Without such an external agency, any agreements between Alice and Bob must be *self-policing*. That is to say, it must be optimal for a player to keep the agreement provided that the other does so too. It follows that only Nash equilibria are available as viable deals. So the only possibility for rational players in the one-shot Trust Minigame is the inefficient outcome in which Alice doesn't deliver because Bob isn't going to pay. But what happens if the game is repeated indefinitely often?

It always remains a Nash equilibrium in a repeated game to keep playing a Nash equilibrium of the one-shot game, but the folk theorem says that there are always lots more equilibria. If the players are sufficiently patient and the probability that the game will be played at least one more time is sufficiently high, then *any* payoff pair inside the cooperative payoff region is available as a Nash equilibrium outcome – provided it pays both players their minimax values or more.

The basic argument is so easy that it is no wonder it was discovered by pretty much everyone who gave repeated games any thought after Nash published his equilibrium idea in 1951. Pick any potential agreement inside the cooperative payoff region. To make this into a Nash equilibrium outcome, we need only punish any players who deviate from the strategy they need to follow in order to implement the agreement. For the purposes of the argument, it is easiest to use the kind of implacable punishment characteristic of the GRIM strategy, in which any deviation is punished forever in the most severe manner available.

What is the most severe punishment Alice can inflict on Bob? The worst she can do is to hold him to his minimax payoff – because he will respond to her attempt to minimize his payoff by making the reply that maximizes his payoff given her choice of punishment strategy. In the Prisoner's Dilemma, the minimax payoff for both players is 0. In the Trust Minigame, Alice's minimax payoff is 1 and Bob's is 0. The heavily shaded regions in Figure 20 therefore show all self-policing agreements for the indefinitely repeated versions of the Prisoner's Dilemma and the Trust Minigame. No other agreements can be supported as Nash equilibrium outcomes.

What can go wrong?

Easy though it is to prove, I think the folk theorem embodies perhaps the most significant insight available to political philosophy. It says that we don't need an external enforcement agency – whether real or invented – to cooperate successfully. In a repeated situation, we can enjoy all the fruits of cooperation by acting as our own policemen.

However, the folk theorem has a major limitation. It assumes that any deviations from equilibrium will be observed by the other players. This is probably not a bad assumption in the case of the small bands of hunter-gatherers in which human culture first evolved. As in the small towns of today, presumably everybody knew everybody else's business. But this certainly isn't true of modern city life. In the anonymity of a big city, it isn't possible to detect and punish deviants often enough to deter cheating. We do our best with CCTV cameras, policemen, auditors, tax inspectors, and the like, but nobody would want to claim that our efforts in this direction are anywhere near efficient.

I wish I could say that game theorists have all the answers to the imperfect monitoring problem, but it remains largely *terra incognita* in spite of the efforts of many clever people. This is perhaps the area of game theory in which further progress would have the greatest social benefits.

Tit-for-tat

Most social scientists think that everything they need to know about reciprocity is summarized in the strategy TIT-FOR-TAT for the indefinitely repeated Prisoner's Dilemma. It tells a player to begin by playing *dove* and then to copy whatever the other player did last time. It is a Nash equilibrium if both Alice and Bob play TIT-FOR-TAT just as it is a Nash equilibrium if they both play GRIM, but TIT-FOR-TAT doesn't punish a deviation relentlessly. Repentant sinners are forgiven after they return to the fold by playing *dove* again.

The popularity of TIT-FOR-TAT derives from Bob Axelrod's Olympiad, in which social scientists were invited to submit computer programs to be matched against each other in the indefinitely repeated Prisoner's Dilemma. After learning the outcome of a pilot round, the contestants submitted programs that implemented 63 of the infinite number of possible strategies for the game.

The most successful strategy in the competition was TIT-FOR-TAT. So Axelrod continued by simulating the effect of evolution operating on all 63 strategies. The fact that TIT-FOR-TAT was the most numerous of all the surviving programs at the end of the evolutionary simulation clinched the question for Axelrod, who then proceeded to propose TIT-FOR-TAT as a suitable paradigm for human cooperation across the board. In describing its virtues, he says:

> What accounts for TIT-FOR-TAT's robust success is its combination of being nice, retaliatory, forgiving and clear. Its niceness prevents it from getting into unnecessary trouble. Its retaliation discourages the other side from persisting whenever defection is tried. Its forgiveness helps restore mutual cooperation. And its clarity makes it intelligible to the other player, thereby eliciting long-term cooperation.

But to describe TIT-FOR-TAT as the most successful strategy in Axelrod's simulation was to gild the lily. Six of the strategies entered for Olympiad survived the evolutionary process, and so the true winner was the mixed strategy in which the surviving strategies are played with the frequency with which they were present when the process stabilized. The frequency of TIT-FOR-TAT in this mixture of six strategies was actually only a little more than one-sixth. Nor is the limited success TIT-FOR-TAT enjoyed in the simulation robust when the initial population of entries is varied. The unforgiving GRIM does extremely well when the initial population of entries isn't biased in favour of TIT-FOR-TAT.

Axelrod defined a nice strategy to be one that is never the first to play *hawk*, but it isn't true that we can count on evolution to generate nice behaviour as he suggests. When some small fraction of suckers worth exploiting is allowed to flow continually into the system, mean strategies outperform TIT-FOR-TAT. The simplest such mean strategy is TAT-FOR-TIT, which begins by playing *hawk* and thereafter switches its action if and only if the opponent played *hawk* last time. Two TAT-FOR-TIT strategies are a Nash equilibrium in the infinitely repeated Prisoner's Dilemma, in which cooperation is achieved only after the first round of play.

As for clarity, it is only necessary for cooperation to evolve that a mutant be able to recognize a copy of itself.

All that is then left on Axelrod's list is the requirement that a successful strategy be retaliatory. This is perhaps the claim that has done most harm, because it applies only in *pairwise* interactions. For example, it is said that reciprocity can't explain the evolution of friendship. It is true that the offensive–defensive alliances of chimpanzees can't be explained with a tit-for-tat story. If Alice needs help because she is hurt or sick, her allies have no incentive to come to her aid, because she is now unlikely to be useful as an ally in the future. Any threat she makes to withdraw her cooperation will therefore be empty. But it needn't be the injured party who punishes a cheater in multi-person interactions. Others will be looking on if Bob abandons Alice to her fate, and they will punish his faithlessness by refusing to form alliances with him in the future. After all, who wants to make an alliance with someone with a reputation for abandoning friends when they are in trouble?

I think the enthusiasm for TIT-FOR-TAT survives for the same reason that people used to claim that it is rational to cooperate in the one-shot Prisoner's Dilemma. They want to believe that human beings are essentially nice. But the real lesson to be learned from Axelrod's Olympiad and many later evolutionary simulations is infinitely more reassuring. Although Axelrod's claims for TIT-FOR-TAT are overblown, his conclusion that evolution is likely to generate a cooperative outcome seems to be genuinely robust. We therefore don't need to pretend that we are all Dr Jekylls in order to explain how we manage to get on with each other fairly well much of the time. Even a society of Mr Hydes can eventually learn to coordinate on an efficient equilibrium in an indefinitely repeated game.

Emergent phenomena

Game theory models of social relationships are sometimes criticized as reductionist because they make no reference to notions like authority, blame, courtesy, duty, envy, friendship, guilt, honour, integrity, justice, loyalty, modesty, ownership, pride, reputation, status, trust, virtue, and the like. The inference is that game theory is an inhuman discipline which treats people like robots.

It is true that – like all successful sciences – game theory is reductive, but it doesn't follow that game theorists think that concepts like authority or duty are irrelevant to human behaviour. On the contrary, we believe that such notions are emergent phenomena that arise when people try to make sense of the equilibria they find themselves playing in the game of life.

For example, the folk explanation of the equilibrium in which Alice always delivers and Bob always pays in the Trust Minigame is that Bob can't afford to lose his reputation for honesty by cheating on Alice, because she will then refuse to provide any service to him in the future. In practice, Bob will usually be someone new, but the same equilibrium works just as well, because nobody will be any more ready than Alice to trade with someone with a reputation for not paying.

Far from denying such stories, game theory offers a nuts-and-bolts explanation of why they sometimes work – and why they sometimes fail. For example, our critics say that we are wrong about the Trust Minigame, because people still pay up, even in one-shot games in which their reputation for honesty is irrelevant. But I notice that filling stations increasingly make you pay in advance for your petrol, presumably because they have experienced the subgame-perfect equilibrium in the one-shot Trust Minigame too often to be willing to play that game any more.

Authority

David Hume tells us that the authority of popes, presidents, kings, judges, policemen, and the like is just a matter of convention and habit. Alice obeys the king because such is the custom – and the custom survives because the king will order Bob to punish Alice if she fails to obey. But why does Bob obey the order to punish Alice? In brief, who guards the guardians?

Game theory answers this ancient question by showing that a version of the folk theorem holds not only for Nash equilibria, but for subgame-perfect equilibria as well. When such an equilibrium is in use, it is always *optimal* to punish any deviant behaviour that leads us to a subgame off the equilibrium path. If you deviate yourself by trying to evade the cost of punishing a deviant, you will take us to another subgame where it is optimal for some other player to punish you. If he fails to do so, we go to yet another subgame – and so on forever.

Immanuel Kant naively thought that to contemplate such chains of responsibility is to initiate an infinite regress, but the folk theorem shows that the chains of responsibility can be bent back on each other. With only a finite number of players, these chains of responsibility are necessarily closed in a manner that Kant failed to consider. Alice obeys the king because she fears Bob will otherwise punish her. Bob would obey the order to punish Alice because he fears that Carol will otherwise punish him. Carol would obey the order to punish Bob because she fears that Alice will otherwise punish her.

At first sight, such a spiral of self-confirming beliefs seems too fragile to support anything solid. It is true that the beliefs go round in a circle, but the folk theorem shows that their fragility is an illusion, since the behaviour generated by the beliefs holds together as a subgame-perfect equilibrium.

Duty

Anthropologists tell us that pure hunter-gatherer societies have no authority structure. Food is gathered and distributed on the principle that all contribute according to their ability, and benefit according to their need.

How can such a social contract survive? If only the tit-for-tat mechanism were available, why would anybody share food with powerless folk outside their family? But the punishment for failing to share needn't be administered by whoever is left to go hungry. In modern foraging bands, the whole group joins in punishing a deviant.

To see how this can work, imagine a toy world in which only a mother and a daughter are alive at any time. Each player lives for two periods. The first period is her youth, and the second her old age. In her youth, a player bakes two (large) loaves of bread. She then gives birth to a daughter, and immediately grows old. Old players are too feeble to work, and so produce nothing.

One equilibrium requires each player to consume both her loaves of bread in her youth. Everyone will then have to endure a miserable old age, but everyone will be optimizing given the choices of the others. All players would prefer to consume one loaf in their youth and one loaf in their old age. But this 'fair' outcome can only be achieved if the daughters all give one of their two loaves to their mothers, because bread perishes if not consumed when baked.

Mothers can't retaliate if their daughters are selfish, but the fair outcome can nevertheless be sustained as an equilibrium. In this fair equilibrium, a conformist is a player who gives her mother a loaf if and only if her mother was a conformist in her youth. Conformists therefore reward other conformists, and punish nonconformists.

To see why a daughter gives her mother a loaf, suppose that Alice, Beatrice, and Carol are mother, daughter, and granddaughter. If Beatrice neglects Alice, she becomes a nonconformist. Carol therefore punishes Beatrice, to avoid becoming a nonconformist herself. If not, she will be punished by her daughter – and so on. If the first-born player is deemed to be a conformist, it is therefore a subgame-perfect equilibrium for everybody to be a conformist. However, the injured party is never the person who punishes an infringement of the social contract. Indeed, the injured party is dead at the time the infringement is punished!

In real life, we say that daughters have a *duty* to care for their helpless mothers. The model shows how such duties could be honoured in a rational world even if all daughters were stonyhearted egoists.

Role of the emotions

Emotions were once dismissed as irrational urges left over from our evolutionary history. The socially aroused emotions associated with pride, envy, and anger are still counted among the seven deadly sins. But if these emotions are as self-destructive as tradition holds, how come evolution equipped us with them? I share the now widely held view that tradition is plain wrong in seeing no useful role for our emotional reactions to social events.

For example, the prototypical scenario for the expression of anger arises when Alice treats Bob unfairly. In his anger at her injustice, he is then likely to inflict some harm on her. Alice therefore takes care to keep her acquisitive urges under control lest she incur his ire.

In this way, it is possible to sustain efficient equilibria in repeated games without any of the players even being aware that they are playing a repeated game. How else would chimpanzees be capable of sustaining high levels of reciprocal altruism? How would humans be capable of the same feat if we always had to spend half an hour or more calculating what to do before taking any action? Some of our thinking in these situations must surely be hardwired, and perhaps getting into an emotional state is simply how it feels when our autopilot takes over the controls.

Revenge

Suppose Bob risks damage to himself in launching an angry attack on Alice after she has treated him unfairly. His behaviour might then easily be dismissed as irrational by observers who fail to notice that he isn't necessarily acting wildly in a one-shot game, but may be carrying through his part of an equilibrium in an indefinitely repeated game.

Experiments on the Ultimatum Game currently provide a focus for this kind of confusion. Why don't experimental subjects accept anything they are offered in one-shot versions of the Ultimatum Game? A popular answer is that they get angry and refuse out of spite. Analysis of the testosterone levels in their sputum would seem to confirm that this explanation is right insofar as it goes.

But why do responders get angry? I think that they get angry because this is their habituated response to an unfair offer in the situations in which we encounter ultimata in real life. Such behaviour survives in repeated situations because it serves to police an equilibrium. It is triggered in one-shot laboratory games because the subjects don't initially appreciate how the laboratory game differs from the games of life to which they are accustomed. But it doesn't follow that we are mere robots controlled by our emotions. Subjects commonly adapt their behaviour to the one-shot games they play in the laboratory as they gain experience. In the Prisoner's Dilemma, it takes only about ten trials for 90% of subjects to learn that playing *dove* doesn't make sense in a one-shot game.

Chapter 6
Information

In a game of perfect information like Chess, the players always know everything that has happened so far in the game. When information is imperfect, we have to keep track of what the players know as they climb the game tree. Von Neumann taught us to do this using the simple idea of an information set.

Figure 21 shows two ways of expressing a simultaneous-move game like Matching Pennies as a tree with information sets. It doesn't matter who moves first if the player who moves second doesn't know what action the first player has taken, so we can make either Alice or Bob move first. In the case when Alice moves first, we enclose Bob's two decision nodes in an information set to indicate that he doesn't know whether he is at his left node or his right node when he moves.

The more information sets we put into an extensive form, the smaller its strategic form becomes. The reason is that a pure strategy only specifies an action at each of a player's information sets – not at each of the decision nodes it contains.

If we took away the information set in the version of Matching Pennies in which Alice goes first, Bob would have $4 = 2 \times 2$ pure strategies, each of which specifies a choice of one of his two actions for each of Alice's two actions. With the information set, he can't

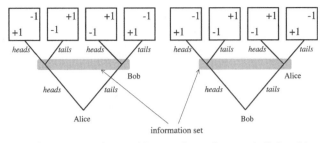

21. Information sets for Matching Pennies. A player can't distinguish between two decision nodes enclosed in the same information set. Alice's payoffs are always in the southwest of each box

make his action contingent on Alice's action, because he isn't informed of her action. So he only has two pure strategies, one for each of his actions.

Poker

The archetypal game of imperfect information is Poker. Unlike Chess, we can solve simple versions of Poker explicitly.

It was Von Neumann's analysis of Poker that made me into a game theorist. I knew that good Poker players bluff a lot, but I just didn't believe that it could be optimal to bluff as much as Von Neumann claimed. I should have known better than to doubt the master! After much painful calculation, I not only had to admit that he was right, but found myself hopelessly hooked on game theory ever after.

However, you are in for a disappointment if you hope to make yourself rich by playing your maximin strategy at the Poker table. Players at the World Poker Championships in Las Vegas play a lot more like Von Neumann recommends than amateurs like you and me, but legendary players like the great Amarillo Slim don't triumph because they play according to the minimax theorem.

22. Full house

Not only would playing your maximin strategy yield an average profit no better than zero in a fair game, it would mostly be as entertaining as watching paint dry.

For example, if Alice were dealt four eights when playing Bob at Straight Poker, her maximin strategy says to reraise four times – but then to fold if Bob raises her again! To make money at a real Poker table, you need to be a lot more enterprising. You must actively seek out and exploit the psychological flaws of your opponents. But unless you are a natural master of human psychology like Amarillo Slim, your naive attempts to exploit the flaws of others are likely to end up with them exploiting yours instead!

Bluffing

Don't worry if you don't know the difference between a straight flush and a full house, or the betting rules in Texas Hold'em. Von Neumann's toy model abstracts all such complications away.

Alice and Bob are each dealt a number between 0 and 1. Both aim to maximize their average dollar earnings on the assumption that all numbers are equally likely to be dealt to your opponent, regardless what is dealt to you. So if Alice is dealt 0.667, she thinks it roughly twice as likely that she has a higher card than Bob.

Before the deal, each player puts an ante of $1 into the pot. After the deal, there is a round of betting, during which Bob may fold. If he folds, Alice wins the pot, no matter who has the better hand. If Bob doesn't fold, there is a showdown, after which the player with the higher card takes the pot. The showdown occurs when Bob calls Alice's bet by making his total contribution to the pot equal to hers.

Von Neumann's model severely restricts the betting possibilities. Alice can first either check (by adding $0 to the pot), or raise (by adding $1 to the pot). If she checks, Bob must call. If Alice raises, Bob has a choice. He can fold or call.

Figure 23 shows the players' maximin strategies in Von Neumann's model. Everybody who plays nickel-and-dime Poker knows that Alice must sometimes raise with poor hands, or Bob

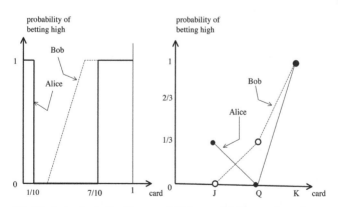

23. Maximin play in Von Neumann's Poker model. The graph on the left shows how the probability with which Alice or Bob should bet high varies with their hands. (Only one of Bob's many optimal strategies is shown.) The graph on the right shows that optimal play in our simplified version has a similar character

will learn never to call when she raises with a good hand. Amateurs try to compromise by bluffing with middle-range hands, but the maximin strategy isn't so timid. If you want to break even at Poker against good opposition, bluff a lot with really bad hands! The point of bluffing is not so much that you might win with a bad hand, as that you want to encourage the opposition to bet with middle-range hands when you have a good hand.

An even simpler model

The next model simplifies Poker even more by replacing Von Neumann's numerical cards by a deck with only the King, Queen, and Jack of Hearts. However, Figure 23 shows that the maximin strategies still look like those in Von Neumann's model.

The chance move that begins the game tree in Figure 24 represents the dealer shuffling the deck into one of six equally likely orders. The top card is then dealt to Alice, and the second card to Bob. The rest of the game tree then shows Von Neumann's betting rules in operation with the new deck of cards.

The game tree looks so formidable that you will probably be surprised to find that you know everything you need to solve the game. First delete dominated strategies by thickening branches that are obviously better than their rivals. For example, Alice should check when holding the Queen, because Bob only calls when he has her beat. Only two decisions then remain in doubt. Does Alice bluff when holding the Jack? Does Bob call when holding the Queen?

Figure 25 shows all the pure strategies of the game, but only the shaded part of the strategic form matters, because the strategies that don't correspond to the shaded part are dominated. The figure also shows a close-up of the shaded part. We can work out the mixed Nash equilibrium of this game by finding what strategies Alice and Bob must use to make their opponent

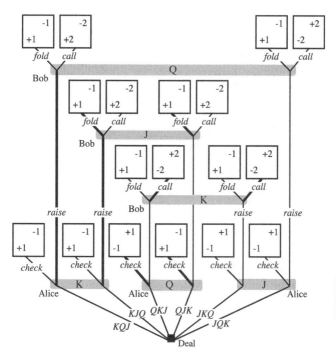

24. Von Neumann's model. After lines have been thickened to show dominant actions, only the cases in which Alice holds the Jack and Bob holds the Queen remain

indifferent. It turns out that Alice should play *RCR* (raise when holding the Jack) with probability 1/3. Bob should play *CCF* (call when holding the Queen) with probability 1/3.

Types

According to the philosopher Hobbes, man is characterized by his strength of body, his passions, his experience, and his reason. In game theory, Alice's strength of body is determined by the rules of the game. Her passions translate into her preferences, and her experience into her beliefs. Her reason leads her to behave rationally.

25. The payoff table for Von Neumann's Poker model. The strategy *RCR* for Alice requires that she raises when holding the King, checks when holding the Queen, and raises when holding the Jack. The strategy *CFF* for Bob requires that he calls when holding the King, folds when holding the Queen, and folds when holding the Jack

These four properties determine a player's type. Unless someone says otherwise, a rational analysis of a game takes for granted that the types of all the players are common knowledge. This very large assumption is sometimes emphasized by saying that information is *complete* in a game.

When is it reasonable to assume that information is complete? A game like Chess creates no problem, but what about Chicken? Is it really likely that Alice will know Bob's degree of risk aversion sufficiently well to work out his payoffs? What will Bob believe about her payoffs? What will Alice believe about what Bob believes about her payoffs?

John Harsanyi

Alice doesn't know Bob's hand in a Poker game. Bob doesn't know what Alice believes about his hand. Alice doesn't know what Bob believes about what she believes about his hand. And so on. We close this seemingly infinite chain of beliefs about beliefs by assuming that the chance move which represents shuffling and dealing the cards is common knowledge. John Harsanyi taught us how to use a similar trick whenever information is incomplete.

Harsanyi had an eventful life for an academic. As a Jew in Hungary, he escaped disaster by the skin of his teeth not once, but twice. Having evaded the death camps of the Nazis, he crossed illegally into Austria with his wife to escape persecution by the Communists who followed. And once in the West, he had to build his career again from scratch, beginning with a factory job in Australia.

As with many truly original minds, his talent was initially unrecognized. It took 25 years for economists to appreciate the ingenuity of his idea for dealing with incomplete information, but he was still alive when they awarded him a Nobel Prize in 1994, alongside John Nash and Reinhard Selten, for his work on incomplete information in games. Whether his equally important work on utilitarianism will ever be adequately recognized now that he is dead remains to be seen.

Incomplete information

Although economists talk about games of incomplete information, there is really no such thing. Harsanyi's theory shows how a situation with incomplete information can sometimes be transformed into a game of imperfect information, which we then analyse using Nash's idea of an equilibrium.

When information is incomplete, the problem is usually that the players may be of various types with different preferences and beliefs. Harsanyi proposed handling this kind of situation as though each player were dealt a type, as in a game of Poker.

The characteristics of such a typecasting move need to be common knowledge if Harsanyi's approach is to work. Economists seem untroubled by this requirement, but my own view is that the method is only genuinely viable when all the knowledge that needs to be held in common is actually available in a database that everybody knows everybody can access.

	slow	speed
slow	3 / 3	? / 0
speed	0 / 4	-1 / -1

one-sided ignorance

	slow	speed
slow	3 / 3	? / 0
speed	0 / ?	-1 / -1

two-sided ignorance

26. Incomplete information in Chicken

One-sided ignorance in Chicken

Middle-aged businessmen often play Chicken when driving oncoming cars in narrow streets. In the version with one-sided ignorance shown in Figure 26, everything about the game is common knowledge except Bob's largest payoff – the payoff he gets from speeding when Alice goes slow.

To apply Harsanyi's method, imagine a chance move that deals Bob his type, which is identified with his top payoff in this example. A pure strategy for Bob then says what action he would take for each type he might be dealt.

The probabilities with which the chance move deals different types to Bob are determined by Alice's beliefs. To tie things down, assume it is common knowledge that Alice believes that Bob's type is equally likely to be anything between 3 and 9. If St Francis of Assisi were driving Bob's car he would doubtless be of type 3, but even Attila the Hun has a type no greater than 9 in this toy model.

The implicit subjunctives built into this modelling technique often create unease. If Bob knows he isn't a saint, why should he behave as though he were playing a game in which he might be St Francis?

The reason that we need Bob to examine how he would have behaved for all possible types that he might have been is that *Alice* doesn't know which of these types has been realized. Since her choice of strategy depends on what Bob would have done if his type were something other than it actually is, Bob can't decide what to do when he knows that his type is $4\frac{1}{2}$ or $5\frac{3}{4}$ without simultaneously considering what he would have done if his type were any of the other possibilities.

Bob learns his type but Alice remains ignorant of the outcome of the typecasting move. She therefore has only two strategies: *slow* and *speed*. Bob has an enormous number of strategies, but we only consider those in which he chooses *slow* if his type is less than some number and *speed* if it is larger.

We have previously located three Nash equilibria for Chicken: two pure and one mixed. The pure equilibria continue to be equilibria in the game with one-sided ignorance. The first equilibrium corresponds to the convention *Ladies' First:* Alice plays *speed* and Bob always plays *slow* whatever his type turns out to be. The second equilibrium corresponds to the convention *Gentlemen First:* Alice plays *slow* and Bob always plays *speed*.

Where no such convention is available – as when driving in heavy traffic – we have to look at an analogue of the mixed equilibrium of Chicken in which each player chooses *slow* and *speed* equally often. We begin by making Alice indifferent between *slow* and *speed*. Any mixed strategy is then optimal for her. Bob can make Alice indifferent by choosing to play *slow* when his type is between 3 and 6, and *speed* when his type is between 6 and 9. It will then seem to Alice that Bob is playing *slow* and *speed* equally often. But Alice mustn't play *slow* and *speed* equally often as in the mixed equilibrium of Chapter 2, since it now has to be optimal for Bob to switch from *slow* to *speed* when his type is 6. To make him indifferent between *slow* and *speed* when his top payoff is 6, Alice must play *speed* three times as often as *slow*.

Notice how the accident rate goes up when we make Alice ignorant of Bob's type. In the mixed equilibrium of the original version of Chicken, Alice and Bob each chose *speed* half the time, and so the probability is 1/4 that both players speed and so cause an accident. In the corresponding equilibrium of the version with one-sided ignorance, the probability of an accident rises to 3/8.

Two-sided ignorance in Chicken

The second example of Figure 26 is more fun than the first, since both Alice and Bob are now ignorant, but the symmetry of the problem makes it even simpler to analyse.

We follow Harsanyi again by introducing a chance move that independently deals a type to each player that the other believes is equally likely to be anything between 3 and 9. The interesting equilibrium arises when Alice and Bob both play *slow* when their type is less than 5 and *speed* when their type is more than 5. It will then seem to both players that their opponent is playing *speed* twice as often as *slow*. A player with a top payoff of 5 will then be indifferent between playing *slow* and *speed*. Optimal play for both Alice and Bob then consists of switching from *slow* to *speed* when their type reaches 5.

The probability of an accident is now 4/9, which is an increase on the probability of 3/8 we found in the case of one-sided ignorance, but we are about to see that increasing the level of ignorance can sometimes make the players better off.

Ignorance is bliss?

Alice and Bob are about to play the Nash equilibrium we just found for Chicken with two-sided ignorance. Each is of type 4, and so both are planning to play *slow*. There is then no possibility of an accident and each player will receive a payoff of 3 utils.

Pandora is a well-informed dogooder who observes that Alice and Bob are basing their choice of strategy on a false premise. Each is behaving as though their opponent might have any type between 3 and 9, but their opponent's type is actually 4. Pandora therefore intervenes with a public announcement that makes it common knowledge that Alice and Bob are both of type 4. Alice and Bob then play the familiar mixed equilibrium for Chicken in which each player chooses *slow* and *speed* half the time. So Pandora's intervention increases the probability of an accident to 1/4 and reduces Alice's and Bob's average payoff to $1\frac{1}{2}$ utils.

Increasing everybody's knowledge can therefore make everybody worse off. Better knowledge is only sure to be an unmitigated good to a player if it is secretly acquired. Thus if Pandora *secretly* tells Alice what the game really is, Alice will switch from *slow* to *speed,* and her payoff will improve from 3 to 4 utils.

The conclusion that revealing information can hurt a society raises an important ethical question. Should it be legitimate for politicians to conceal the truth for our own good? John Stuart Mill is perhaps the most benign of a number of philosophers from Plato onwards who have answered *yes* to this question, but my own view is *no*. I keep my mouth shut when I learn that someone is cheating on their marriage, but I want people to think it right to blow the whistle in public life. Lies that supposedly further the public interest usually turn out to benefit only the liars who tell them.

Signalling your type

When people play Chicken in real life, they look out for clues that may signal the type of their opponent. Is Alice driving a dented old pick-up truck? Is Bob wearing a dog collar?

To be effective in signalling a player's type, a signal must usually be costly to send. If Alice is dealt a pair of twos in Poker, it won't help her to tell Bob that she has been dealt four aces. Game theorists dismiss such idle bombast as *cheap talk*. Bob will only pay attention if Alice puts her money where her mouth is. But if she bluffs by betting like she has four aces, she risks being called and losing her bet.

Returning to Chicken with two-sided ignorance, suppose that both Alice and Bob can simultaneously send a costly signal that says, 'I am a high type – don't mess with me.' If some types send this signal to register their strength, then remaining silent becomes a signal of weakness. We therefore need to contemplate a new game in which both players have a strategic opportunity to signal strength or weakness before playing Chicken. We look at a particular subgame-perfect equilibrium in which players whose type exceeds 5 signal strength, and players whose type is less than 5 signal weakness.

If Alice claims to be strong by sending the signal and Bob implicitly admits to being weak by remaining silent, the equilibrium requires that they play Chicken according to the rule: Ladies First. That is to say, Alice speeds and Bob goes slow. If Bob sends the signal and Alice doesn't, they play according to the rule: Gentlemen First. Bob then speeds and Alice goes slow.

The more interesting cases arise when either both or neither send the signal. We have already seen how to analyse the versions of Chicken with two-sided ignorance that result. In the case when both send a signal, it becomes common knowledge that both types lie between 5 and 9. It is then a Nash equilibrium if players whose type is less than 6 choose *slow* and players whose type is more than 6 choose *speed*. In the case when neither sends a signal, it becomes common knowledge that both types lie between 3 and 5. It is then a Nash equilibrium if players whose type is less than 4 choose *slow* and players whose type is more than 4 choose *speed*.

It is only at the final step that any new considerations arise. It needs to be optimal for players whose type is more than 5 to send the signal and for players whose type is less than 5 to remain silent. A player whose type is exactly 5 will then be indifferent between sending the signal and remaining silent. We therefore need to work out what a player of type 5 expects to get in the two cases. For the equilibrium to hold together, the cost of claiming to be tough must be equal to the difference between these two payoffs. The difference turns out to be $2\frac{1}{6}$ utils, and so the cost of a signal must also be $2\frac{1}{6}$ utils if the equilibrium is to work.

Displays

Costly signalling is of major importance in bargaining. Delay is frequently used for this purpose. For example, during a strike, complaints are sometimes made about the irrationality of union officials who don't plan to consider the latest offer from the employer until Tuesday week. Union officials are doubtless sometimes irrational, but exactly the same tactic might be used by rational players sending a costly signal of their strength. In its crudest form, a bargainer might rationally display his strength simply by burning a bundle of bank notes.

Biology offers wonderful examples. Avishag Zahavi describes costly signalling as the Handicap Principle. Why do some skylarks sing when pursued by a hawk? To signal that they are swift enough to escape even when they handicap themselves by singing. Young hawks still give chase, but they soon learn not to bother. Why do peacocks have such magnificent tails? Part of the reason is runaway sexual selection. Peahens like big tails and so peacocks with big tails father more chicks. But big tails presumably began as a costly signal of a peacock's fitness.

Chapter 7
Auctions

Alice now becomes a boss who wants her subordinates to work towards her aims rather than pursuing their own interests. In the language of economics, Alice is a principal and the subordinates are her agents.

Alice could tell her agents what to do under all possible contingencies, but there are two reasons why such command economies are notoriously inefficient. The first is that it is difficult for a principal to monitor her agents to ensure that they are following her rules rather than doing their own thing. The second reason is that the agents often know their own business better than the principal.

Mechanism design

The rules that Alice can monitor and enforce create a game for the agents to play. To persuade the agents to pursue her aims rather than their own in situations that she is unable to monitor or lacks the expertise to regulate, she needs to invent suitable incentives to motivate the agents. The problem of finding a good system of regulations and incentives is called *mechanism design*.

The chief insight from game theory is that people should be expected to change their behaviour after a new reform is introduced. Their behaviour will adjust until they eventually settle on a Nash equilibrium of the new game. When Alice evaluates a possible new mechanism, she should therefore ask herself how much she likes what will happen *after* the agents have moved to an equilibrium of the new game. The almost universal mistake that real principals make is to ask instead how much they like what will happen *before* the agents learn the ropes of the new game.

In a real-life example, the new chairman of the controlling body of a university health scheme argued in favour of abolishing its co-pay arrangements. These require you to pay the first hundred pounds or so of any claim you make, with a view to discouraging frivolous use of the service. To make up the lost revenue, he proposed that the premiums be increased by enough to cover the co-pay receipts from the previous year. When the economist on the committee objected that the premiums would need to be increased by more than this, a vote was taken on whether anyone else thought that 'people would visit their doctor when they didn't need to'. Only the economist voted *yes* to this loaded question, but there wasn't enough money to pay the bills in the following year.

The United States Congress made a bigger mistake in 1990 when it passed an act intended to ensure that Medicare wouldn't pay substantially more for its drugs than private health providers. The basic provision of the act said that a drug must be sold to Medicare at no more than 88% of the average selling price. The problem was created by an extra provision which said that Medicare must also be offered at least as good a price as any retailer. This provision would only work as its framers intended if drug manufacturers could be relied upon to ignore the new incentives created for them by the act. But why would drug manufacturers ever sell a drug to a retailer at less than 88% of the current average price if the consequence is that they must then sell the drug at the same price to a huge customer like Medicare? However, if no drugs are sold at less than 88% of the current average, then the average price will be forced up!

In 2006, the British Liberal Democrats proposed introducing green taxes that would allow income tax to be reduced by a total of $12 billion. This proposal failed not only to appreciate that people change their behaviour in response to new taxes, but that the very purpose of a green tax is to change behaviour!

Nobody would ever propose constructing an aeroplane or a bridge without giving a great deal of thought to how the mechanism would stand up to the stresses and strains it will face when built, but the idea that one should give the same care and attention to the design of social mechanisms is typically greeted with scorn. I once provoked outright laughter when I suggested that some money might be spent testing a major reform in a psychological laboratory to see whether it would work before being put into practice. Even the design of big-money auctions is still often left in the hands of amateurs who know nothing whatever of the simple ideas to be outlined in this chapter.

Judgement of Solomon

When confronted by two women disputing the motherhood of a baby, King Solomon famously proposed that the baby be split in two, so each claimant could have half. The false mother accepted the judgement, but the true mother's 'bowels yearned upon her son' so that she begged for the baby go to her rival rather than being hacked in two. Solomon then knew the true mother, and awarded her the baby.

Actually, the biblical story doesn't support Solomon's proverbial claim to wisdom particularly well. His scheme would have failed if the false mother had been more strategically minded. So what scheme would work better?

Solomon is the principal. The plaintiff and the defendant are the agents. Trudy is the true mother, and Fanny is the false mother. To keep things simple, we assume it to be common knowledge that

Trudy would pay all she has in the world for her baby, but Fanny will pay only some lesser amount.

Solomon's aim is to award the baby to the true mother, but he doesn't know what type each agent is. He could ask them, but Fanny has no incentive to tell the truth. So Solomon follows Harsanyi's methodology by imagining a chance move that either casts Trudy in the role of the plaintiff and Fanny in the role of the defendant, or Trudy in the role of the defendant and Fanny in the role of the plaintiff.

Figure 27 shows the rules of a mechanism that achieves the first-best outcome of awarding the baby to the true mother for certain. The plaintiff moves first by saying whether or not she claims to be the mother. If she denies being the mother, the baby is given to the defendant. If she claims to be the mother, the defendant must say whether or not she claims to be the mother. If she denies being the mother, the baby is given to the plaintiff.

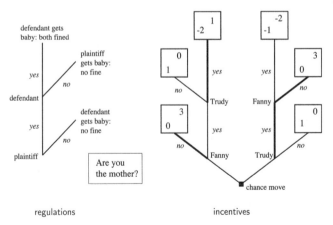

regulations

incentives

27. The Judgement of Solomon. The diagram on the right shows the game from Solomon's point of view when Trudy values the baby at 3 shekels, Fanny values the baby at 1 shekel, and Solomon sets a fine of 2 shekels

If both women claim to be the mother, the baby is given to the defendant, and both women are fined.

Solomon must use his famed wisdom in setting the incentives for Fanny and Trudy. The fine must be higher than Fanny's valuation of the baby but lower than Trudy's. The thickened lines in Figure 27 then show the result of applying backward induction. When the actors use this subgame-perfect equilibrium, Trudy *always* gets the baby, and no fine is paid.

Economic applications

The big successes of mechanism design are in auctions and regulatory economics. The billions of dollars raised in telecom auctions have attracted a lot of attention, but the regulatory applications are probably of more significance in the long run.

The fat cats who get regulated squeal a great deal about the virtues of the free market, but they know that the pleasant properties of perfectly competitive markets only apply when there are large numbers of small buyers and sellers. When there are only small numbers of sellers, they always end up using their market power to screw the consumer unless restrained by government regulation.

The free marketeers are right that no regulation is usually better than the bad regulation with which we have mostly been saddled in the past, but regulation doesn't need to be bad. Better regulation using the principles of mechanism design is already making headway against the opposition of the various gurus whose advice it makes redundant. However, I don't have space to talk seriously about regulatory economics here, and so the rest of this chapter will be about auctions.

Willam Vickrey

Auctioning is the branch of game theory in which most progress has been made. It is also an area in which game theory has been used with spectacular success to solve applied problems. The amount of money raised in telecom auctions designed by game theorists is astronomical.

Once upon a time, governments used to organize what became known as 'beauty contests' when allocating valuable public assets to private companies. Each company would submit a mighty document explaining why it should get the asset rather than one of its rivals. A committee of officials would then decide whose document they liked best. But the officials commonly had no idea of the commercial value of the assets they were selling. Nor were they told the truth by the contestants in the beauty contest. Why would agents ever tell the truth to their principal unless it were to their advantage? They need to be offered appropriate incentives before they will part with the information the principal needs. Auctions persuade agents to tell the truth is by making them put their money where their mouths are.

William Vickrey is the hero of auction theory. He was advocating the use of specially designed auctions for the sale of major public assets 30 years before the idea became popular. The Federal Communications Commission ignored his arguments until the word finally reached Congress, which insisted that the next batch of telecom frequencies be sold by auction. A design by a group of game theorists led by Paul Milgrom then raised a total of $20 billion in revenue.

The team I led myself designed other telecom auctions in Britain, Belgium, Denmark, Greece, Hong Kong, and Israel. The auction in Britain raised $35 billion all by itself. After the collapse of the NASDAQ index in 2001 and the consequent bursting of the hitech bubble, there was a great deal of caterwauling as telecom executives sought to blame their own failure to assess the market properly on the game theorists who supposedly made them pay more for their licences than they were worth. But who but an idiot would bid more for something than they thought it was worth?

I think all the brouhaha only served to underline how effective game theorists can be when allowed to apply the discipline of mechanism design on a large scale. The Nobel Committee obviously thought the same thing and belatedly awarded Vickrey a Nobel Prize, but he died three days after receiving the announcement.

Kinds of auction

Auctions are nothing new. Herodotus describes the auctioning of unwanted wives in ancient Babylon. Nor are big-money auctions a novelty. In AD 193 the Pretorian Guard auctioned off the Roman Empire to one Didius Julianus. Some of the kinds of auctions in regular use are listed below.

English auctions

Sotheby's uses this kind of auction to sell old masters. An auctioneer invites oral bids. The bidding continues until nobody is willing to bid any more. The auctioneer traditionally cries out,

'Going, going, gone!'

28. Going, Going, Gone!

If nobody interrupts with a new bid, he hammers his block, and the object is sold to the buyer who bid last.

Dutch auctions

The auctioneer begins by announcing a high price, which is then gradually lowered until a buyer calls a halt. The first buyer to do so then acquires the object at the price outstanding when he or she intervened.

Dutch auctions can be quick, and so are used to sell perishable goods like fish or cut flowers. In Amsterdam's flower auction, a seller may fly cut flowers in from Zambia, and the buyer may ship them out to sell them in Chicago all in a single day. However, slow-motion Dutch auctions are sometimes operated by used furniture stores that reduce the price of unsold items by 10% each month.

First-price, sealed-bid auctions

This is the standard format for government tenders. Each potential buyer privately writes his bid on a piece of paper and seals it in an envelope. The seller commits herself to selling the object to whoever makes the highest bid at the price he bid.

Vickrey auctions

In a Vickrey auction, the object is sold by the sealed-bid mechanism to the highest bidder, but at the highest price bid by a *loser*. This will be the *second*-highest price unless there is a tie for first place, in which case the winner is chosen at random from the highest bidders.

Milton Friedman led an irresponsible but successful campaign to persuade the American government to change the auction format used to sell Treasury Bonds to what financial gurus choose to call a second-price auction, but he was mistaken in supposing that he was advocating the appropriate generalization of a Vickrey auction to the case when many identical objects are for sale. This is only one example of a distressing phenomenon. Even in those few cases where game theorists know the answer to a problem, government officials usually prefer the advice of false prophets with better public relations.

Optimal auctions

Alice wants to sell her house, which is worth nothing to her if a buyer can't be found. The only potential buyers are Bob and Carol. What auction design should Alice use?

Alice's problem is similar to King Solomon's. She doesn't know Bob and Carol's valuations for her house. If she knew their valuations, she would simply make a take-it-or-leave-it offer of one penny less than the higher of the two valuations. It would then be rational for the bidder with the higher valuation to accept, because a profit of one penny is better than nothing. This analysis depends heavily on Alice having the power to make an irrevocable commitment to the rules of the Ultimatum Game. Mechanism design doesn't work at all if the agents don't believe that the principal will stick by the rules of the game she invents for them to play.

Since Alice doesn't know Bob and Carol's valuations, she follows Harsanyi's methodology by modelling her uncertainty about their valuations with the help of a chance move. In the simplest case, this chance move assigns valuations to Bob and Carol independently – so that nothing you may learn about one agent's valuation tells you anything new about the valuation of the other agent. Where it is necessary to be specific, I shall assume that each agent's valuation for Alice's splendid house is equally likely to be anything between zero and 36 million dollars.

Most people are surprised to discover that all the auction designs surveyed in the previous section are optimal for Alice if everybody is trying to maximize their average profit in dollars. Alice should set a reserve price of 18 million dollars and then use whichever of the designs takes her fancy, because they all yield the same average profit of 15 million dollars.

Reserve prices

Notice that Bob won't bid at all half the time, because his valuation will be less than Alice's reserve price. Since the same is true of Carol, Alice won't sell her house at all one time in every four. If this happens, she mustn't cheat on her own rules by auctioning the house again with a lower reserve price – unless she doesn't care whether the agents will believe her in the future when she acts as a principal again!

I have given advice to many governments on designing big-money auctions, but I have never had any success in persuading the responsible officials to set as high a reserve price as I wished. Nor have I ever had any confidence that the officials understood that they mustn't put the object for sale back on the market immediately if it didn't reach its reserve price. But my efforts on this front aren't entirely wasted, since they provide an excuse for looking only at the case of a zero reserve price in what follows. All the auction designs we have considered so far raise the same revenue on average whatever the reserve price, but it is easiest to explain why in the case when all the potential buyers can be counted on to make a bid.

Revenue equivalence

In a symmetric Nash equilibrium, agents with higher valuations will make higher bids in all of the auctions we have considered. So the agent with the higher valuation will win. The probability of an agent with a given valuation winning the auction is therefore the same for all our designs.

What does the agent expect to pay on average? In each of our auctions, the answer turns out to be *half* the agent's valuation.

It follows that Alice's average revenue must be the same for each design. This average revenue turns out to be only 12 million dollars, but this is the best Alice can hope for if Bob and Carol bid rationally and she sets a reserve price of zero.

Why should the winner expect to pay only half his or her valuation? The case of an English auction is easiest. It is obvious that Bob and Carol should each keep bidding until the going price reaches their valuation, after which they should shut up. So the bidding will stop when the going price reaches the lower of the agents' valuations. The winner will therefore pay the *loser's* valuation. Since the loser's valuation is equally likely to be anything less than the winner's valuation, its average value is half the winner's valuation.

It now becomes possible to see why a Vickrey auction is essentially the same as an English auction. If Bob and Carol submit their true valuations to Alice and she sells her house to the highest bidder at a price equal to second-highest bid, then her average profit will be exactly the same as in an English auction. But why should Alice expect Bob and Carol to bid their true valuations?

The answer is that this strategy dominates all the agents' other alternatives. If the opposing agent has bid less than your valuation, you make sure of winning without affecting the price you pay by bidding your valuation. If the other agent has bid more than your valuation, you don't want to win and so you might as well bid your valuation.

What about a Dutch auction? Bob and Carol could write the price at which they plan to stop a Dutch auction on a piece of paper and leave Alice to implement the strategy on their behalf. A Dutch auction is therefore essentially the same as a first-price sealed-bid auction. So if we can work out what price Bob and Carol should seal in their envelopes in the latter auction, we will also know when they should plan to stop a Dutch auction.

Shading your bid

What bid should you seal in your envelope in a first-price sealed-bid auction? You certainly won't bid your true valuation, because your profit would then be zero if you won. You need to shade your bid down. But by how much? I already gave the secret away when explaining why all our auction designs are revenue equivalent. Bob and Carol should each bid only half their valuations.

This fact is usually proved by calculus but I plan to explain it using landscape gardening. What shape should a garden be if you want to enclose the maximum area with a fixed amount of fence? The answer is obviously a circle. If the garden has to be rectangular, the answer is a square.

Returning to the problem of finding a symmetric Nash equilibrium for a first-price, sealed-bid auction, imagine it to be common knowledge that Carol will make a bid proportional to her valuation. Bob's probability of winning will then be proportional to whatever bid he makes. He wants to maximize his average profit, which is the product of his profit if he wins and his probability of winning. This product is proportional to the area of a rectangular garden whose length is his profit if he wins and whose width is his bid.

Since his profit if he wins is the difference between his valuation and his bid, the length of the garden fence is proportional to his valuation. The fence therefore has the same length whatever bid he makes. So finding Bob's best reply reduces to deciding which rectangular garden has the largest area when enclosed by a fence of fixed length. Since the answer is a square, Bob best reply is found by setting his bid equal to the difference between his valuation and his bid – which makes his optimal bid equal to half his valuation. His bid will then be proportional to his valuation, and so Carol's optimal bid will also be equal to half her valuation.

All-pay auctions

Instructors in game theory courses are fond of auctioning a dollar according to the following rules. The bidding is as in an English auction, with the highest bidder getting the dollar, but *everyone* pays their highest bid *including* those who don't win the dollar. Watching the expression on students' faces when the bidding reaches one dollar, and the losers realize that it is now worth their while to bid *more* than one dollar can be quite entertaining!

Bribing corrupt politicians or judges is rather like an all-pay, sealed-bid auction. Everyone pays, but only one bribe can be successful. If there is honour among thieves, it will be largest bribe that carries the day.

All-pay auctions are mentioned here only to make the point that the revenue equivalence result applies very generally. The agents will obviously shade their bids much more in an all-pay, sealed-bid auction than in a first-price, sealed-bid auction, but Alice's average revenue will still be 12 million dollars.

Bespoke auctions

Economic correspondents of newspapers sometimes quote revenue equivalence results when wrongly arguing that it *never* matters what kind of auction the seller uses. But the result goes away if Bob and Carol are budget constrained or risk averse. Alice then gets more in a Dutch auction than an English auction. Nor does it hold if Bob and Carol's valuations cease to be independent. Alice will then expect more in an English auction than a Dutch auction.

When a big-money auction is to be run, it is therefore of paramount importance that its rules be tailored to the circumstances. For example, if Alice somehow knows that the only amounts at which it is possible for Bob and Carol to value her house are 27 million or 36 million dollars, then she shouldn't run any of the auctions considered so far. If the two possibilities are equally likely, she should run a sealed-bid auction in which the winner pays the average of the two bids!

Winner's curse

Agents with private values know their valuations before the auction begins, and nothing they might learn during the auction will lead them to change their valuations. All the auctions we have looked at up to now have been treated as private-value auctions. At the opposite end of the spectrum are common-value auctions, in which it is common knowledge that the value of the object being sold is the *same* for all the prospective buyers.

For example, when licences to drill for oil in undersea tracts are auctioned, the amount of oil in a tract is the same for everybody, but the buyers' estimates of how much oil is likely to be in a tract will depend on their differing geological surveys. Such surveys aren't only expensive, but notoriously unreliable. Some prospective buyers will therefore receive optimistic surveys, and others will receive pessimistic surveys. So who will win the auction?

If Bob treats his survey's estimate of the value of the tract as a private value, then he will win whenever his survey is the most optimistic. But when Bob realizes that his winning the auction implies that all the other surveys are more pessimistic than his, he will curse his bad luck at winning! If he had known at the outset that all the other surveys were more pessimistic than his, he wouldn't have bid so high.

As with the all-pay auction, game theory instructors like to catch their students out by trapping them with a common-value auction. A glass jar filled with coins and rumpled bills of various denominations is auctioned off to the highest bidder, who usually falls prey to the winner's curse and so makes a substantial loss.

Wallet Game

To avoid the winner's curse, bidders must take account of the information that the bids of their rivals convey about how much they think the object for sale is worth. The Wallet Game is a toy example. Alice confiscates Bob and Carol's wallets, and then uses an English auction to sell their combined contents back to them.

It is a Nash equilibrium if both agents plan to keep bidding until the going price reaches twice the money in their own wallets. If Carol bids this way, then Bob will be cursed if he wins by bidding more than twice the money in his own wallet. He then knows that the amount of money in his own wallet is less than half the final price. He also knows that the amount of money in Carol's wallet is less than half the final price, because she stopped bidding. So the total amount of money Bob has won must be less than the price he paid for it.

Chapter 8
Evolutionary biology

Biologists have a huge advantage over social scientists in applying game theory, because they have so much more data. Natural selection has generated a vast variety of different species, some of which are so weird and wonderful that they seem to defy rational explanation. But what could be more satisfying than finally to learn why the unusual genetics of the *Hymenoptera* explain the unequal sex ratio in certain species of bees? Or why two very different variants of the bluegill sunfish succeed in coexisting together in the same lake? To deny evolution in the face of such examples seems to me like emulating the theologian who refused to look through Galileo's telescope.

Even more remarkable is the fact that even the crudest of toy games will sometimes suffice to model some animal behaviour successfully. Nobody really believes, for example, that reproduction is sexless among birds, or that the evolutionary process is deterministic. But, as in physics, the models that result from making such heroic simplifications sometimes fit the data remarkably well.

Evolutionary game theory

Herbert Spencer summarized Darwin's theory of evolution as the *survival of the fittest*. When asked why the animals of some species behave as they do, we therefore look for an answer that explains why alternative behavioural traits were less fit. But how is fitness to be defined?

Bill Hamilton's definition makes it inevitable that modelling animal behaviour will sometimes reduce to finding the Nash equilibria of games. He took the fitness of a behavioural trait to be the average number of extra children that carry that trait into the next generation as a result of the trait being used in the current generation. With this definition, behavioural traits can be identified with strategies, and fitness with utility.

When animals compete, we can then imagine that Chance occasionally picks two or more individuals from the relevant populations to play a game. A famous example in ecology is the predator-prey game that results in the numbers of Canadian lynxes and hares cycling indefinitely. However, this chapter will focus on games played within a single species that have stable outcomes. For example, what determines how long a male dung fly will wait at a particular cow pat for a female when seeking to mate? Since the strategic problem is the same for all dung flies, we can then confine our attention to symmetric Nash equilibria of symmetric games.

A symmetric game looks exactly the same to all its players. In a symmetric equilibrium, all players use the same strategy. A variant of Nash's theorem shows that all finite, symmetric games have at least one symmetric Nash equilibrium.

Replicators

Unfortunately, the philosophical waters have been muddied by a controversy over who or what should be treated as a player in an evolutionary game. A whole species? An individual animal? A package of genetic material? Or an individual gene? The title of

Richard Dawkins' *Selfish Gene* seems to tell us where he stands on this issue, but he actually takes the more sophisticated view that anything which replicates itself may be regarded as the fundamental unit in an evolutionary game.

Like the old lady I once heard giving Dawkins a hard time for failing to see that genes are just molecules and so can't have free will, people often find it paradoxical that game theory could be successfully applied in evolutionary biology. How can an insect be a player in a game? Insects can't reason. Their behaviour is largely instinctive. They just do what they are programmed to do.

The solution to the paradox is that the players in the game needn't be taken to be the animals under study. If the behaviour being investigated is largely instinctive, then it is coded for in the animal's genes. One may think of the genes as part of the hardware of a natural computer: the part where the computer's programs are stored. Some of the programs control the animal's behaviour.

An important property of computer programs is that they can be copied from one computer to another. Computer viruses copy *themselves* from one computer to another. They are *self-replicating*. The programs imprinted on an animal's genes are also self-replicating. But their replication is immensely complicated compared with the replication of a computer virus. Nature not only has to copy programs from one natural computer to another, she has to create a new natural computer to which the programs may be copied. Crick and Watson's discovery of how Nature works this trick using the device of the double helix is one of the great scientific adventure stories. But its thrills will have to be enjoyed elsewhere. All that is important here is that we understand that *something* exists that does two things:

- It replicates itself.
- It determines strategic behaviour in a game.

Whenever we find something in a model on which we can hang these two properties, it will be called a *replicator*.

Genes can certainly be replicators. Critics sometimes complain that a mutation in a single gene is unlikely to have much effect, but even the slightest modification in a behavioural trait can be significant when fitness is averaged out over a long enough time span. Packages of genes that tend to be replicated together also count as replicators. In a parthogenetic species like the mason beetle, a mother transmits its entire genetic coding to its children, in which case one might as well say that each individual type of animal is a replicator.

To survive, replicators need hosts in whose genes they are imprinted. Since we have defined the fitness of a host to be a measure of how frequently it reproduces its genes, it becomes almost a tautology that replicators that confer high fitness on their hosts will come to control a larger share of hosts than those that confer low fitness. If the environment will only support a restricted number of hosts, the replicator conferring low fitness on its hosts may eventually die out altogether. The fittest replicator will then have survived.

If Alice is watching the situation evolve, she might try to make sense of what she sees by attributing a goal or purpose to whatever mechanism generates replicators: that of maximizing the fitness of their hosts. If natural selection operates for long enough in a stable environment, only those replicators that are good at maximizing the fitness of their hosts will survive. To Alice, it will therefore seem *as though* something were consciously choosing replicators to maximize fitness. We call this imaginary *something* a player of the game.

For example, when the replicators are taken to be variants of a single gene, we can imagine the player sitting at the locus on the chromosome where that particular gene operates. Careful biologists who like to think of genes themselves as players use the term *allele* for the possible forms a gene may take. However, it is common to blur the distinction between a player and a replicator in much the same way that the distinction between a player and a type often gets blurred in the theory of incomplete information (see Incomplete information Chapter 6).

Evolutionary stability

For evolution to work, there must be some variation in a population. The shuffling of genes in sexual reproduction is one source of variation. Geographical migration and mutation are others. When can we expect a population to settle down in the face of such random variation? One approach is to look for an asymptotic attractor – a population of replicators that is stable in the face of any small perturbation.

The simplest possible model of a biological evolutionary process is called the replicator dynamics. Figure 14 on P. 50 shows how it works in a particular game when the players are drawn from two different populations that evolve separately. In this chapter, the corresponding diagrams are much simpler, because the focus will be on symmetric games in which the players are drawn from a single population.

The replicator dynamics assumes that the proportion of a population hosting a particular replicator increases at a rate proportional to two factors:

- The fraction of the population currently hosting the replicator.
- The difference between the current fitness of the replicator's hosts and the average fitness of all the hosts in the population.

The first requirement simply recognizes that the rate of growth of a replicator is constrained by the number of parents available to transmit the replicator to the next generation. The second requirement recognizes that evolution can only take account of a replicator's fitness *relative* to the fitness of the population as a whole.

If all the replicators to be considered are present when the replicator dynamics get started, then the system can only converge on a symmetric Nash equilibrium – if it converges on anything at all.

Evolutionarily stable strategies

The idea of an evolutionarily stable strategy or ESS begins with George Price, who submitted a 60-page essay on evolutionary mathematics to the journal *Nature,* which a more worldly author would have known publishes only short articles. Fortunately, his referee was John Maynard Smith. Together they wrote a paper which distilled Price's essential wisdom into something readable. Maynard Smith went on to write *Evolution and the Theory of Games,* which put evolutionary game theory on the map. George Price eventually committed suicide, reportedly because he found it increasingly difficult to reconcile his fundamental contributions to evolutionary biology with his religious convictions.

When the players are drawn from different populations, the considerations to which we are about to appeal lead to nothing more exotic than a strict Nash equilibrium (in which there are no alternative best replies to the equilibrium strategies). But evolutionarily stable strategies apply when the players are drawn from the *same* population to play a symmetric game. The defining properties are:

- An ESS must be a best reply to itself.
- If an ESS isn't the only best reply to itself, it must be a better reply to any alternative than the alternative is to itself.

The first requirement says that a pair of evolutionarily stable strategies must be a refinement of the notion of a symmetric Nash equilibrium. But if this were our only requirement, what would prevent a destabilizing invasion of the population by an alternative best reply? The second requirement provides the necessary evolutionary pressure against such an invasion by asking that an ESS be more fit than an invader immediately after an invasion.

Any ESS in a symmetric game is necessarily an asymptotic attractor of the replicator dynamics. In its turn, an asymptotic attractor is necessarily a symmetric Nash equilibrium. We therefore have a necessary condition and a sufficient condition for evolutionary stability. Both conditions apply to a wider class of evolutionary processes than just the replicator dynamics, but some care is necessary when applying the ESS concept even with the replicator dynamics. For example, the trajectories of the replicator dynamics in Rock-Scissors-Paper go round in circles, and the game has no ESS at all (see Finding maximin strategies, Chapter 2). Worse still, other 3×3 symmetric games exist which have isolated asymptotic attractors that aren't ESS. Only in symmetric games with just two pure strategies is the ESS concept entirely safe.

However, carping about the inadequacies of evolutionary modelling concepts in the abstract isn't very productive. The real question is: how useful are they in making sense of real biological examples?

Hawk-Dove Game

Two birds drawn from the same species occasionally contest some valuable resource. The two replicators in the population make their hosts either passive or aggressive in such situations. A passive bird surrenders the entire resource to an aggressive bird. Two passive birds share the resource equally. Two aggressive birds fight.

Maynard Smith referred to passive birds as *doves* and aggressive birds as *hawks,* for which reason the game is called the

Hawk-Dove Game – but don't be misled into supposing that the birds are intended as representatives of different populations that evolve separately. The environment is intended to be entirely symmetric.

Prisoner's Dilemma

If possession of the resource enhances a bird's fitness by four utils and getting into a fight by only one util, then the Hawk-Dove Game reduces to the version of the Prisoner's Dilemma shown in Figure 29. Recall that the only Nash equilibrium is for both players to use *hawk* (see Chapter 1). Since this strategy is strictly dominant, it is also an evolutionarily stable strategy.

The birds playing the Hawk-Dove Game are drawn from a single population, and so the replicator dynamics for the Prisoner's Dilemma in Figure 29 on P. 126 are one-dimensional (rather than two-dimensional as in previous examples). The arrow shows that there is a unique asymptotic attractor in which the population consists entirely of hawks. If we were to perturb this population by throwing in a positive fraction of dovelike mutants, they would eventually all be eliminated. In fact, the basin of attraction consists of all population states other than that in which the whole population consists of doves. The appearance of even a tiny fraction of hawk mutants therefore dooms the doves to eventual extinction.

Group selection fallacy

The ardor with which game theorists deny the various fallacies claiming that cooperation is rational in the Prisoner's Dilemma pales into insignificance when compared with the almost demonic ferocity with which evolutionary biologists denounce the group selection fallacy.

According to the group selection fallacy, evolution favours mutations that enhance the fitness of the species rather than the fitness of the mutated gene itself. A population playing *dove* in the Prisoner's Dilemma would then be invulnerable to invasion by a mutant *hawk* because any fraction of hawks in the population would reduce the total fitness of the population. The fallacy lies in misplacing the relevant replicator at the level of the species. It is, after all, at the molecular level that replication takes place physically when the double helix divides. It is therefore right to confine attention to the unique ESS, which is *hawk*.

Charles Darwin knew nothing of modern genetics and so occasionally fell into a number of errors, of which the group selection fallacy was one. However, it is the biologist Vero Wynne-Edwards who is the luckless target of modern critics. He suggested, for example, that starlings gather in large numbers at nightfall in order to estimate their numbers with a view to controlling their population size. George Williams's critique of his group selection arguments was very influential, leading to a literature explosion of which Dawkins's *Selfish Gene* is just one example.

The sex ratio problem is a nice example of the failure of the group selection fallacy. What sex ratio would favour a new species? The answer is a few males and many females. So why do we have roughly equal numbers of boys and girls? Because a boy born into a population consisting mostly of girls will be fitter than a girl, and a girl born into a population consisting mostly of boys will be fitter than a boy. How is an equilibrium achieved in such situations? That is our next topic.

Chicken

The payoff values that identify the Hawk-Dove Game with the Prisoner's Dilemma are unrealistic, because even a slight injury is likely to be a serious handicap. If we assign a negative utility to a bird that gets into a fight by subtracting two more utils from its payoff, we are led to the version of Chicken of Figure 29.

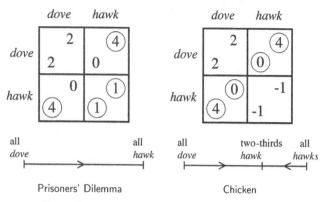

29. Replicator dynamics in the Hawk-Dove Game. For a single population, the trajectories can be shown using one-dimensional diagrams. A similar diagram for the Stag-Hunt Game would show the arrows pointing away from its mixed equilibrium

Chicken has three Nash equilibria: two pure and one mixed. The mixed equilibrium requires that each player use *dove* one-third of the time and *hawk* two-thirds of the time. In contrast to Getting to equilibrium in Chapter 2, we reject the pure equilibria because they are asymmetric – only the symmetric mixed equilibrium is of interest here.

Must we then envisage birds as throwing dice? Not in the Hawk-Dove Game. To see why, imagine that one-third of the population are doves and two-thirds are hawks. When Chance selects two birds at random to play Chicken from this polymorphic population, it will seem to both players that their opponent is playing the mixed equilibrium strategy. Since players are indifferent between the pure strategies they use in a mixed equilibrium (see Chapter 2), both hawks and doves will therefore be equally fit. So no individual bird need randomize at all to sustain the equilibrium.

Maynard Smith observed that the mixed population state not only corresponds to a symmetric Nash equilibrium, it also corresponds to an evolutionarily stable strategy. The replicator dynamics shown in Figure 29 confirm that a population with twice as many hawks as doves is an asymptotic attractor. It is good to be a hawk when there are many doves, and good to be a dove when there are many hawks. The two effects cancel out at the mixed Nash equilibrium, where a player is indifferent between choosing *dove* and *hawk*.

At one time the survival of two variants of the same animal in the same environment was thought to be a mystery. Surely one type must be just a little less fit than the other, and so be driven to extinction in the long run? But the fitness of a variant varies with its frequency in such situations.

Bluegill sunfish

The bluegill sunfish (*Leoponis macrochirus*) is a popular target for anglers in North American lakes. The fish enjoy a synchronized mating season in which males stake out nests on the lake floor. Females lay their eggs in whatever nest takes their fancy. The eggs are promptly fertilized by the resident male who then stoutly guards the resulting brood.

Guardian males share their lakes with another type of male commonly called a sneaker. Sneakers mature after two years, whereas guardians take seven years. Sneakers are unable to establish and guard a nest, since they are little more than self-propelled sexual organs. When a female lays her eggs in a nest, they rush out from their hiding places and try to fertilize her eggs before the guardian.

In a profligate display of exuberence, Nature has also gifted us with an alternative sneaker that disguises itself as a female, and an alternative guardian that establishes its nest at a distance from the closely packed nests established by regular guardians.

The theory says that the numbers of each type of male will adjust to equalize their fitness – a conclusion that is reassuringly in line with the data.

Playing the field

Animals don't need to randomize when they compete in pairs, but they sometimes behave as though they do when 'playing the field'.

Male dung flies (*Scatophaga stercoraria*) congregate at cow pats hoping to mate with females attracted by the scent. How long should a male wait before seeking a fresh cow pat (which takes an average of four minutes)? Game theory tells us to look for symmetric Nash equilibria. In the simplest model, each male will use a mixed strategy in which his waiting time is exponentially distributed. This means that his probability of leaving now is always twice what it would be if he waited some fixed time period. But how long is this period?

If the theory is right, the period should adjust until a dung fly's mating success is the same no matter how long he waits. As with sunfish, the data are very supportive of this hypothesis, although I dare say a dung fly would be no more receptive than a company executive to the idea that he was actively randomizing (see Does randomizing make sense, Chapter 2).

Kin selection

The animal kingdom overflows with examples of cooperation within the family. African hunting dogs regurgitate food to help out a hungry pack brother. Marmosets and tamarins help to care for their nephews and nieces in extended families. Male birds of some species do the same when their chances of being able to reproduce in the current year are low. Aphids give up their lives defending their siblings from attack. Musk oxen form a defensive ring around the weaker members of the family when attacked by wolves. Why is kinship so important in the animal kingdom?

Hamilton's rule

Bill Hamilton's *Narrow Roads of Geneland* is an account of the life and work of another oddball genius. He recently died a typically adventurous death on a field trip to Brazil. Hamilton deserves most of the credit for introducing game theory into biology, although I doubt that he ever heard of John Nash during the long years he strove, alone and unrecognized, to create a whole new field of research. One of his many achievements was to formulate the evolutionary explanation of cooperation within the family nowadays known as kin selection.

His point was famously anticipated in a semi-serious joke of J. B. S. Haldane. When asked whether he would give his life for another, he replied that the sacrifice would only be worthwhile if it saved two brothers or eight cousins! Haldane's joke is only funny if you know that your degree of relationship to a full brother is one-half, and your degree of relationship to a full cousin is one-eighth.

It is sometimes said that the degree of relationship can't really matter, because human beings share nearly all their genes anyway. But this is to miss the point that we are never concerned with genes that are *always* present in the human body, but only with a particular piece of behaviour that will be modified or left alone according to whether a recently mutated gene is present or absent.

Your degree of relationship to a relative is the probability that a recently mutated gene in your body is also in your relative's body. To see that your degree of relationship to a cousin is one-eighth, imagine that your cousin is the daughter of your mother's sister. The probability that a mutant gene in your body came from your mother rather than your father is one-half. If it came from your mother, the probability it is also in the body of your aunt is one-half. If it is in the body of your aunt, the probability she passed it to your cousin is one-half. Multiplying these three halves together, we get one-eighth.

What counts in calculating the fitness of a gene is the average number of times it gets replicated in the next generation. But it doesn't matter which of two or more identical versions of a gene is copied. A copy made from a gene in my sister's body is just as good as a copy made from an identical gene in my own body. When we figure out the fitness of a gene in my body, we therefore have to take account, not only of the effect of my behaviour on my own reproductive success, but of its effect on the reproductive success of my relatives. Hamilton called the outcome of such a calculation an *inclusive* fitness.

If my sister is my only relative, then a mutant gene in my body shouldn't simply count the extra number of children I will have on average as a consequence of its modifying my behaviour. It should use Hamilton's rule, which requires adding in the extra number of children that my sister will have – weighted by one-half, because this is the probability that the mutant gene is also in her body. For example, if I expect to have one child less as a consequence of changing my behaviour, and my sister expects to have four children more, then Hamilton's rule says that the inclusive fitness of my new strategy is $-1 + \frac{1}{2} \times 4 = 1$. My personal loss is therefore outweighed by my sister's gain.

The results of replacing individual fitnesses in a game by inclusive fitnesses can be dramatic, as with the Prisoner's Dilemma in Figure 30. When nestlings compete for food with their siblings, their behaviour is largely genetically programmed. If the nestlings were identical twins, both players could therefore count on their

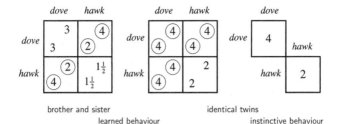

30. Relatives play the Prisoner's Dilemma. Hamilton's rule has been used to convert individual fitnesses in the Prisoner's Dilemma of Figure 29 into inclusive fitnesses. Identical twins will now cooperate, but we must expect mixed behaviour from siblings

opponent choosing exactly the same strategy as themselves. The situation then reduces to a one-player game.

The two-player games of Figure 30 are more suitable for modelling human behaviour in cases when the players' behaviour is learned rather than genetically programmed. When Alice and Bob are identical twins, we are led to a game that I call the Prisoner's Delight because *dove* is now a dominant strategy. When they are brother and sister, we are led to a form of Chicken, in which we should expect both dovelike and hawkish behaviour to survive together. (It would take us too far afield to examine the more interesting case of siblings whose behaviour is genetically determined.)

Kinship is especially important in primitive human societies. In societies that tolerate promiscuity, for example, some of the functions undertaken by a father in our own society are taken on by a child's maternal uncle – the underlying reason being that everybody knows his degree of relationship to the child is one-fourth, whereas nobody can be sure who the child's real father may be. Hamilton's rule provides an evolutionary explanation of such phenomena by quantifying the extent to which we should expect blood to be thicker than water.

Social insects

A species is eusocial if it lives in colonies with overlapping generations in which one or a few individuals produce all the offspring, and the rest serve as sterile helpers. Eusociality is rare except among the *Hymenoptera* – the order of insects that includes ants, bees, and wasps. It used to be said that true eusociality has evolved independently at least 12 times in the *Hymenoptera*, but only twice elsewhere – the exceptional cases being the termites of the order *Isoptera* and the naked mole-rats of the order *Rodentia*. Later research has found other eusoscial species, the most interesting of which is a shrimp (*Decapoda*) that colonizes sponges on coral reefs. But the frequency of eusociality in the *Hymenoptera* remains a puzzle demanding an explanation.

Why did evolution generate casts of sterile workers? Why do they work tirelessly for the sake of others? Why is this phenomenon common among the *Hymenoptera* and rare elsewhere?

At one level, the puzzle is easy. Groups working together are usually more productive than individuals acting separately. In a beehive or an anthill, very large numbers of sterile workers specialize in protecting and caring for the young, while the queen specializes in being an egg-laying machine. As a consequence, the total number of young produced is immensely larger than if pairs of bees or ants brought up separate families by themselves.

It is clear why the queen benefits, but what's in it for the workers? Each fertile child the queen produces is related to the workers. They are the workers' brothers and sisters. A mutant gene that expresses itself in the body of a worker therefore has a lot of relatives to count when it computes its inclusive fitness. All the queen's fertile children – weighted by their degree of relationship to a worker – must be counted when calculating the benefit to a worker of striving hard in support of the queen. The productivity of a beehive or an anthill then ensures that the balance comes down very firmly on the side of eusociality.

All this would be equally true of the human species if we had a sterile worker cast, but we traditionally bring up our children in extended families rather than biological factories. So why didn't evolution take us down the same road as the *Hymenoptera?*

Bill Hamilton's answer to this question depends on the fact that the *Hymenoptera* are haplodiploid, which means that unfertilized eggs grow into haploid males and fertilized eggs grow into diploid females. In a haploid species, each locus on a chromosome hosts just one gene. Humans are diploid, with each locus hosting two genes, one from the mother and one from the father. This is why the degree of relationship between human sisters is one-half, since a child gets one gene from each parent at every locus, and the gene it gets from each parent is equally likely to be either of the two genes the parent carries at that locus. By contrast, the degree of relationship between sisters in the *Hymenoptera* is three-quarters, because each locus on their chromosomes gets the *same* gene from their father, and a randomly chosen gene from the pair carried at that locus by their mother.

Workers therefore have a stronger motivation to help their fertile sisters than humans would have in the same situation. But this isn't the end of the story.

Robert Trivers pointed out that the degree of relationship between genetically female workers and their brothers, the drones, is only one-fourth. If the sex ratio among the *Hymenoptera* were 50 : 50, then the average degree of relationship between a sterile worker and a fertile sibling would therefore only be the average of three-quarters and one-quarter, which is one-half – the same as in our own species. However, in some species among the *Hymenoptera*, the sex ratio is about 75 : 25 in favour of fertile females as opposed to fertile males. How come? The answer is not only of interest in itself, but it also serves to complete Hamilton's explanation of why eusociality has evolved so often in the *Hymenoptera*.

In the *Hymenoptera*, it is usually the genes expressed in the workers that determine the sex ratio, because it is the workers who rear the young. The sex ratio must therefore make a worker indifferent between raising a fertile male or a fertile female infant. This happens only when the sex ratio is $75:25$, because the payoff to a mutant gene expressed in the body of a worker is then $\frac{3}{4} \times \frac{1}{4}$ from producing a male and $\frac{1}{4} \times \frac{3}{4}$ from producing a female. Since these payoffs are equal, a mixed equilibrium can survive in which females are born with probability three-quarters and males with probability one-quarter.

With this sex ratio, the average degree of relationship a sterile worker has with a fertile brother or sister is $\frac{3}{4} \times \frac{3}{4} + \frac{1}{4} \times \frac{1}{4}$, which is five-eighths. If the worker were human, the degree of relationship would be one-half. Human workers would therefore have to work harder on behalf of a human queen in order to derive the same benefit as a worker in an anthill or a beehive.

It should be emphasized that the details of this overly simple story are controversial among biologists. Even those species which come closest to fitting the story deviate in idiosyncratic ways. But I think that the fact that game theory allows evolutionary biologists to explain sex ratios in species where these aren't symmetric is one of the more convincing demonstrations that we must be doing something right.

Of course, many mysteries remain. Why are the *Hymenoptera* haplodiploid? How come only some species in the order are eusocial? What about colonies with multiple queens? What of *Pachycondyla villosa*, in which species *unrelated* queens apparently found colonies together? What of the many puzzles posed by termites? Creationists seize on such admissions of ignorance as an excuse to debunk evolutionary science, but I think their criticisms simply reveal a failure to understand how science works.

Evolution of cooperation

We already know that cooperation can be sustained among animals that aren't related by the mechanism that Bob Trivers called reciprocal altruism. A wonderful example is provided by the vampire bat (*Desmodus rotundis*).

Vampire bats roost together in caves during the day. At night they seek an animal from which to suck blood. Some 8% are unsuccessful, which is a big problem for bats, who need to feed every 60 hours or so. For this reason, the evolutionary pressure towards sharing is very strong. Gerald Wilkinson discovered that vampire bats share blood on a reciprocal basis with roostmates who aren't always relatives. In brief, a bat is more likely to

31. Vampire bat

regurgitate blood for a begging roostmate, if the roostmate has shared blood with it in the past.

How does such cooperation get off the ground? Axelrod has muddied the waters by claiming to have shown that TIT-FOR-TAT is an ESS in the indefinitely repeated Prisoner's Dilemma. Although Maynard Smith mistakenly endorsed the claim, it obviously isn't true. A population of TIT-FOR-TATs can be invaded by the strategy that always plays *dove*. Such a mutant won't displace TIT-FOR-TAT, but nor will it be expelled.

No pure strategy can be an ESS in the indefinitely repeated Prisoner's Dilemma: a mutation that changes the strategy at an unreached subgame won't even be detected, let alone driven out. The ESS concept needs to be widened to be useful in such a setting, so that whole sets of strategies through which a population may drift are regarded as evolutionarily stable aggregates. For example, the set N in both Figure 14 and Figure 32 is a kind of aggregate asymptotic attractor within which the system is free to drift. (There needn't be a trajectory leading away from N, as in both these cases.)

Hawk-Dove-Retaliator Game

The problem is already apparent in the Hawk-Dove-Retaliator Game with which Maynard Smith and Price originally explored the evolution of cooperation. A retaliator plays like a hawk against a hawk, and like a dove against a dove. The *retaliate* strategy is weakly dominated, and so the game has a symmetric Nash equilibrium in which *retaliate* is not played at all. As in the Hawk-Dove game, *dove* is played with probability 1/3 and *hawk* with probability 2/3. In the upper triangle of Figure 32, this mixed equilibrium is marked with the letter M. There are also an infinity of Nash equilibria in which *hawk* is not played at all, marked in Figure 32 with the letter N. These require that *retaliate* is played with probability at least 3/5.

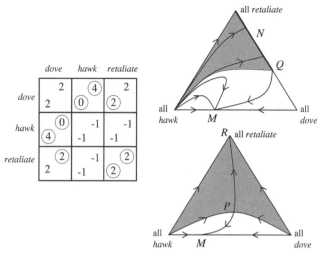

	dove	hawk	retaliate
dove	2 ④ 2 ⓪	④ 2 ⓪ ②	
hawk	⓪ -1 ④ -1	-1 -1	
retaliate	② -1 2 -1	② ②	

32. Hawk-Dove-Retaliator Game. The upper triangle shows the replicator dynamics in the pure game. A population mix is thought of as being distributed at the three corners of the triangle. The point that represents it is located at the centre of gravity of this distribution. The lower triangle shows the replicator dynamics when retaliators have a small advantage over doves, and hawks have a small advantage over retaliators

The upper triangle shows the replicator dynamics for the Hawk-Dove-Retaliator Game. The shaded set is the basin of attraction for the set N. Maynard Smith and Price ignore this set because only M is an ESS. However, if the system found its way into N, its only chance of escaping is if a new *hawk* mutation appears while it is close to Q. But this rare event might be delayed for a very long time. There have, in fact, been enormously long perods of stasis in the evolution of many species that might be attributed to this cause.

The lower triangle of Figure 32 shows the replicator dynamics for a modified version of the Hawk-Dove-Replicator Game in which a retaliator is realistically assumed to do a little better against a dove and a little worse against a hawk. This game has three symmetric Nash equilibria. There is an analogue of the mixed equilibrium M of the Hawk-Dove Game; a pure equilibrium R in which only *retaliate* gets played; and an equilibrium P in which all three strategies are played with positive probability. The equilibria M and R correspond to ESS strategies.

The basin of attraction of R is shaded in Figure 32. Since this is a large set, we have a toy model in which it makes sense to apply the ESS concept, and which offers the beginnings of an explanation of the evolution of cooperation. Maynard Smith and Price expand the model by introducing a bullying type who displays like a retaliator but backs down when challenged. The bullies displace doves, but otherwise nothing much changes.

However, the most interesting application of the Hawk-Dove-Replicator Game is to the case of local interaction. In real life, animals mostly play games with their geographical neighbours. Chance might therefore easily fix it so that a mutant retaliator becomes numerous in a small neighbourhood. The Hawk-Dove-Replicator Game then tells us that the other strategies will gradually be extinguished in that neighbourhood. But then the same will happen in overlapping neighbourhoods until the whole environment is taken over by retaliators.

This seems to me the most convincing toy explanation of the evolution of cooperation that is commonly offered.

Social or cultural evolution

It is sometimes thought that one should only speak of evolution if the analogy with biological evolution is very close. It is true that replicators don't only arise in a biological context.
Rules-of-thumb, codes-of-conduct, fashions, lifestyles, creeds, and scientific ideas are all replicators of a kind. Richard Dawkins refers

to such cultural replicators as memes. They spread from one human mind to another through imitation or education.

I used to be enthusiastic about memes, but now that we understand that the replicator dynamics emerge not just from toy models of biological reproduction, but also from toy models of imitation and stimulus-response learning, it seems unnecessary to shackle ourselves to the meme paradigm. Whenever adaptive dynamics take us to the Nash equilibrium of a game, I am ready to speak of cultural evolution.

The chief difference in applying evolutionary ideas between the biological and the social sciences would seem to be that biologists are usually very well informed about the sources of interesting variation, whereas social scientists can only guess. For example, an evolutionary model in economics must take account of the fact that mutations in the form of new money-making schemes are appearing all the time, but if economists could predict which of these were going to be successful, they would all be rich!

Chapter 9
Bargaining and coalitions

Half of Von Neumann and Morgenstern's *Theory of Games and Economic Behavior* is devoted to two-person zero-sum games. This part of the book is the birthplace of the *noncooperative* theory of games that we have studied up to now. In this theory, the strategic opportunities of the players are explored in detail, and predictions of their behaviour are made using the idea of a Nash equilibrium. The other half of Von Neumann and Morgenstern's book is about *cooperative* game theory.

This use of words is an endless source of confusion, because critics mistakenly assume that noncooperative game theory is exclusively about conflict and cooperative game theory is exclusively about cooperation. They are right to the extent that cooperative game theory is largely about how rational people will cooperate. What coalitions will form? Who will get how much of the gravy? But they go astray when they treat cooperative and noncooperative game theory as antithetic perspectives in which Dr Jekyll and Mr Hyde are set up as rival paradigms on the human condition. After all, the folk theorem is part of noncooperative game theory, but its chief interest lies in showing how cooperation can be sustained as equilibrium behaviour in repeated games.

Cooperative game theory differs from noncooperative game theory only in abandoning any pretension at explaining *why* cooperation survives in our species. It postulates instead that the players have access to an unmodelled black box whose contents somehow resolve all the problems of commitment and trust that have worried us periodically throughout this book. Among other things, the black box must contain an explanation of how preplay negotiations on how a game should be played can result in an agreement that the players treat as unconditionally binding.

In economic applications, one can sometimes argue that the black box contains all the apparatus of the legal system. The players then honour their contracts for fear of being sued if they don't. In social applications, the black box may contain the reasons why the players care about the effect that behaving dishonestly in the present may have on their reputation for trustworthy behaviour in the future. One can even argue that the black box contains the results of our childhood conditioning, or an inborn aversion to immoral behaviour.

The utopian fallacy is to imagine that the black box of cooperative game theory contains nothing more than the fond hope that conflict would disappear if only people would behave rationally. Much conflict in real life is admittedly stupid, but we won't make people less stupid by teaching them that their hearts are more rational than their heads.

The way to respond to the utopian fallacy is to open the cooperative black box and take a long hard look at what lies inside. Why does it make sense for players to trust each other in some situations and not in others? Why don't they pursue their own interests rather than those of the group to which they belong?

When seeking to answer such questions, we have no choice but to use the methods of noncooperative game theory. Noncooperative game theory is therefore the study of games in which any cooperation that may emerge is fully explained by the choice of strategies the players make. But this can be very hard. Cooperative game theory bypasses all the difficult *why* questions in the hope of finding simple characterizations of *what* agreement rational players will eventually reach.

Nash program

The *Nash program* invites us to open the cooperative black box to see whether the mechanism inside really does work in the way a particular cooperative solution concept assumes.

Nash observed that any negotiation is itself a species of game, in which the moves are everything the players may say or do while bargaining. If we model any bargaining that precedes the play of a game in this way, the result is an enlarged game. A strategy for this negotiation game first tells a player how to conduct the preplay negotiations, and then how to play the original game depending on the outcome of the negotiations.

Negotiation games must be studied without presupposing preplay bargaining, all preplay activity having already been built into their rules. Analysing them is therefore a task for noncooperative game theory. So we look for their Nash equilibria, hoping that the equilibrium selection problem won't prove too difficult.

When negotiation games can be solved successfully, we have a way of checking up on cooperative game theory. If a cooperative solution concept predicts the result of a rational agreement on how to play some game, then a noncooperative analysis of the enlarged negotiation game should yield the same answer.

Nash therefore regarded cooperative and noncooperative game theory as complementary ways of approaching the same problem. Cooperative game theory offers easily applied predictions of rational agreements. Noncooperative game theory provides a way of testing these predictions.

Nash bargaining solution

A Beverly Hills mansion is worth $4m to its owner and $5m to a potential buyer. By getting together and agreeing a sale, the buyer and the seller can create a surplus of $1m. How this surplus is divided between them is decided by bargaining. A simple model that captures the essence of this archetypal bargaining problem is traditionally known as Divide-the-Dollar.

The story that goes with the model envisages a philanthropist who offers Alice and Bob the opportunity to share a dollar – provided they can agree on how to divide it between them. If they can't come to an agreement, the philanthropist takes his dollar back again. In this story, the dollar represents the surplus over which two economic agents bargain. The philanthropist's provision that the dollar is available only if Alice and Bob can reach an agreement represents the fact that there will be no surplus unless the agents get together to create it.

When Nash considered the problem, orthodox economists held that rationality is irrelevant to the problem, because the outcome depends on how skilfully Alice and Bob negotiate. Bargaining was therefore thought to be a problem for psychology rather than economics. Even Von Neumann and Morgenstern endorsed this view in their *Theory of Games and Economic Behavior*. When speaking about bargaining 30 years later, I found that hecklers were still sold on the idea that 'bargaining isn't part of economics'. In retrospect, it seems amazing that such a bizarre notion should have won such widespread acceptance, but the past is truly a foreign country.

Nash's argument

John Nash began to think about bargaining when he took a single economics course on international trade as part of his undergraduate degree. The thoughts to which he was led

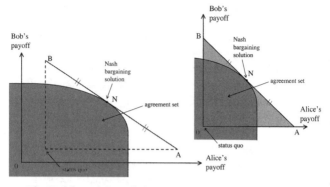

33. The Nash bargaining solution

eventually overturned the orthodox view that the bargaining problem is indeterminate.

Recall that we can identify utility with money in the case of a risk-neutral player. If Alice and Bob are risk neutral in Divide-the-Dollar, one therefore doesn't need to be a genius to predict that they will split the dollar fifty-fifty if they both have access to the same strategic opportunities in whatever negotiation game they play. But suppose they have different attitudes to risk? If Bob is more risk averse than Alice, he will fear a disagreement more than her. So he will end up with less of the dollar – but how much less?

Nash's way of figuring out the answer is illustrated in Figure 33. The first step is to identify each available deal with the pair of utilities that Alice and Bob would get if the deal were implemented. The disagreement point that results if no deal is reached at all is called the *status quo*. The shape of the set of all possible agreements is convex when Alice and Bob are both risk averse.

Nash assumed that a rational deal will be somewhere on the frontier of the set of all possible agreements – otherwise the players would be able to find a more efficient agreement that both prefer. (Economists mysteriously call this observation the Coase theorem, although it is neither a theorem nor original to the Nobel Prize-winner Ronald Coase.) Nash then drew a tangent to the frontier of the agreement set at the point representing the rational deal.

The next step is to relocate the zeros on Alice and Bob's utility scales at the *status quo*. The units on their utility scales are then adjusted to make the slope of the tangent at the rational deal equal to 45°. The diagram on the right of Figure 33 shows the new situation. If the agreement set were the shaded triangle in this diagram, then the rational deal would have to be the midpoint of the hypotenuse (for the same reason that we agreed that the rational deal would be fifty-fifty in Divide-the-Dollar when Alice and Bob are risk neutral).

The final step is to argue that the rational deal must remain the same when we throw away all the points in the shaded triangle on the right of Figure 33 that aren't in the deeply shaded agreement set. Nash called these 'irrelevant alternatives' because Alice and Bob didn't choose any of them when they were available, and so they have no reason to change their agreement when they cease to be available.

To find the Nash bargaining solution in the diagram on the left of Figure 33, we therefore need only locate the tangent that touches the frontier of the agreement set at its midpoint

Who should do how much housework?

Newspapers like to stoke the gender wars when short of copy. Here is a typical quote: 'Men pay lip service to equal rights in the home while letting women do three quarters of the household chores.' Other things being equal, the fact that wives do more housework than husbands would indeed show that the balance of power within marriages is biased in favour of men, but are other things equal?

Alice and Bob are getting married. They have no interest in enjoying any of the benefits of marriage other than sharing the housework. In the modern style, they agree on a binding marriage contract that specifies how many hours a week of housework each will contribute. What deal does the Nash bargaining solution predict that they will reach?

In a toy version of the problem, Alice thinks a household should devote two hours a day to housework; Bob thinks one hour a day is adequate. Each player derives a benefit of 100 utils a week if at least the number of hours they think appropriate is worked; otherwise they see no benefit at all in any housework being done.

Neither Alice nor Bob likes doing housework. Alice loses 5 utils a week for each hour of housework that she does. Bob loses 10 utils per hour, because he dislikes doing housework more than Alice. In the *status quo* situation before the marriage, Alice therefore does 14 hours of housework a week from which she derives a utility of 30 utils; Bob does 7 hours of housework from which he also derives a utility of 30 utils.

The Coase theorem says that the bargaining outcome will be efficient, which means that Alice will get her way over the number of hours that the new household will spend on housework. To find the Nash bargaining solution, we need to find the extreme outcomes that just make the marriage worthwhile for both partners. One extreme arises when Alice does all the housework; she will then get 30 utils and Bob will get 100 utils. The other extreme arises when it is Bob who gets only 30 utils. He will then do one hour of housework a day. Alice must do the other hour of housework to make up the two hours a day she thinks necessary. Her utility will then be 65 utils.

Because the model has been fixed to make Alice and Bob risk neutral, the Nash bargaining solution is found by averaging the two extremes. So Alice will end up with 47.5 utils and Bob with 65 utils a week. To make this happen, Alice will have to work $10\frac{1}{2}$ hours a week and Bob only $3\frac{1}{2}$ hours a week.

The Nash bargaining solution therefore says that if Alice and Bob bargain on an equal basis, then Alice will get her way on the number of hours worked a week, but she will have to do three-quarters of the work. If it is indeed true that wives do three times as much housework as single women, then our toy model shows that it doesn't necessarily follow that the balance of power within marriages is biased in favour of men. Who would do how much housework if all the factors left out of the toy model were taken into account? Even if I knew, I wouldn't say!

Rubinstein's bargaining model

In accordance with the Nash program, Nash defended his bargaining solution with a noncooperative bargaining model in which Alice and Bob each simultaneously commit themselves to take-it-or-leave-it demands. However, Schelling was later successful in casting doubt on the realism of attributing commitment power to the players in negotiation games.

For example, if Bob were able beat Alice to the draw when making an irrevocable commitment in Divide-the-Dollar, then he could scoop the pot by demanding 99 cents, leaving Alice with a choice between one penny or nothing. But how does Bob convince Alice that he is truly committed – that nothing she might do can make him revise his demand? Who believes someone who claims he is now making his 'last and final offer'? Even prices posted on expensive items in fancy stores are seldom final. The seller will try to make you feel like a cheapskate for challenging the price, but folk wisdom is right for once. Everything is negotiable. Never take no for an answer.

It is genuinely hard to establish commitments. People sometimes make a career of building up a reputation for being stubborn or stupid for this purpose. Trade unionists occasionally succeed in committing themselves by voting for intransigent leaders. But outside such special circumstances, the vocabulary of commitment is usually just so much cheap talk. But if all threats must be credible, we have seen that we need to look at subgame-perfect equilibria.

So what happens when anything a player says has to be credible before the other player will believe it? This question led Ariel Rubinstein to make the most important of all contributions to the Nash program. In the most natural noncooperative model of bargaining, Alice and Bob alternate in making offers to each other until they reach agreement. If they are assumed to prefer making any particular deal now rather than later, then Rubinstein showed that the alternating-offers model has a unique subgame-perfect equilibrium.

My own contribution was to show that the unique subgame-perfect equilibrium outcome approximates an asymmetric version of the Nash bargaining solution when the time interval between successive offers becomes sufficiently small. In the symmetric version of the Nash bargaining solution, the ratio NB/AN in Figure 33 is equal to one. In the asymmetric version NB/AN equals the ratio of the rates at which Alice and Bob discount time.

If we make Alice more patient than Bob, her discount rate will decrease and so the asymmetric version of the Nash bargaining solution predicts that she will get more of the surplus available for division.

What matters in bargaining?

When I first met Ariel Rubinstein, he told me that he had been working on the bargaining problem without any success. Since his theorem proved to be pivotal in destroying the orthodoxy that the bargaining problem is indeterminate, this was an unduly humble judgement. But his reason for judging himself so harshly still holds good. All the work on bargaining that has been summarized up to now assumes that information is complete. But how often do real bargainers know each other's preferences?

When Alice tries to sell Bob a house, she would like to know the most that he would really be willing to pay – but he won't tell her. Nor will she tell him the lowest price she will take. Such informational asymmetries matter enormously. In particular, the Coase theorem fails. Roger Myerson has shown that, if it is common knowledge that Alice and Bob's valuations are independent and equally likely to be anything between \$4m and \$5m, then the result of optimal bargaining is very inefficient indeed. Even when the bargaining process is chosen to maximize the expected surplus that rational bargainers can achieve, the house is sold only when it is worth \$250,000 more to Bob than to Alice!

Information trumps all other considerations, but nobody knows how to extract a unique prediction from Rubinstein's bargaining model when information is incomplete. The following principles therefore have solid foundations only when the bargainers are unable to conceal any secrets from each other.

Commitment

It is in Alice's interest to convince Bob that she is unable to accept less than her current demand, but she should view any attempt by him to claim to have made a similar commitment with severe scepticism. Sometimes your opponent will even turn out to be a pushover. For example, when I once asked a car-hire clerk what discounts were offered on the price she had just quoted, she replied: 20%.

Risk

The players' attitudes to taking risks determines the shape of the agreement set. The more risk averse you are, the less you get. Used car dealers therefore affect light-hearted unconcern at the prospect of losing a sale. But as the Bible tells, although they say it is naught, when they goeth their way, they boasteth.

Time

The relative patience of the players determines how much asymmetry must be built into the Nash bargaining solution. The more impatient you are, the less you will get. In a recent court case in which I was involved, the telecom regulator had ruled that the leading fixed-line company must do a deal with a new entrant on the price it would charge to connect the entrant's customers with its own customers. But in the absence of a ruling on *when* the deal must be in place, the fixed-line company could afford to be infinitely patient and thereby expropriate all the gains from trade.

Playing fair?

A best-selling book on negotiation dismisses any use of strategy as a dirty trick. One should supposedly just insist on what is fair. It is perhaps for this reason that a study of collective wage bargaining in Sweden came up with 24 different definitions of what counts as fair!

A reputation for religious or moral obstinacy can certainly sometimes be strategically useful when bargaining. For example, Daniel Defoe's *Compleat English Tradesman* explains that it was contrary to the religious principles of the Quakers of his time to 'lie' by asking for a deal better than they were willing to take. They were therefore successful bargainers, because their first offer was known to be their final offer. But what if the opposition attempts the same commitment tactic? War is the usual consequence when two nations put rationality aside in this kind of way.

None of this is meant to imply that fairness is somehow irrational. On the contrary, it seems to me to be the most important of the conventions that humans use to resolve equilibrium selection problems in everyday coordination games. But rather than regarding fairness as a substitute for compromises reached by rational bargaining, John Rawls's *Theory of Justice* makes rational bargaining the foundation stone of his definition of a fair outcome. Rawls identifies a fair deal with the agreement that Alice and Bob would reach if they were to bargain behind a 'veil of ignorance' that concealed their identity during the bargaining. Neither Alice nor Bob would then wish to disadvantage anyone, because they might themselves turn out to be the disadvantaged party.

I have devoted a substantial chunk of my life using game theory to examine the implications of Rawls's definition. Why does it strike us as reasonable? Does it lead to a utilitarian outcome as claimed by Harsanyi, or an egalitarian outcome as claimed by Rawls? However, life is too short to explain why I think Rawls defended a sound intuition with a wrong argument.

Forming coalitions

How can we apply what we have learned about how two rational people bargain to the bargaining that takes place when coalitions form? Von Neumann and Morgenstern proposed the simplest toy model in which coalitions matter.

Alice, Bob, and Carol are to play Divide-the-Dollar. Who gets how much is determined by majority voting. Any coalition of two players can therefore dispose of the dollar as they choose. But which coalition will form? Who will be the odd man out? How will the dollar be divided?

Outside options

Alice's outside option when bargaining with Bob is the most she can get elsewhere if their negotiations break down altogether.

Labour economists continue to make the error of identifying the *status quo* payoffs with the players' outside options when using the Nash bargaining solution to predict the outcome of wage negotiations. For example, if Bob will become unemployed if he fails to come to an agreement with Alice, then his *status quo* payoff is taken to be the level of social benefit.

To see why it is usually a mistake to use the Nash bargaining solution in this way, it is necessary to modify the Rubinstein bargaining model so that Alice and Bob always have the opportunity to take up their outside option after refusing an offer. It then becomes obvious that the outside options are relevant to the bargaining outcome only to the extent that we should discard all payoff pairs from the agreement set that assign somebody less than their outside option. The *status quo* needs to be identified with the payoffs the players receive *while* negotiating. For example, if Alice and Bob are seeking to negotiate the end of a strike, then their *status quo* payoffs are their respective incomes *during* the strike.

In order for it to be right to identify the *status quo* payoffs with the players' outside options, any breakdown in the negotiations needs be forced rather than voluntary. To model such a forced breakdown in Rubinstein's model, one can introduce a chance move that ends the negotiations with some small probability after each refusal. This would correspond to the case in which any delay in reaching an agreement might result in the surplus over which Alice and Bob are bargaining being stolen by a third party.

Odd-Man-Out

Our three-player version of Divide-the-Dollar can be regarded as three two-player bargaining problems to which we can apply Nash's cooperative bargaining theory. When two players bargain about how they will split the dollar should they agree to form a coalition on how to vote, their outside options are the deals that each would reach if they were to bargain with the odd-man-out instead.

It follows that Alice must expect the same payoff if she succeeds in forming a coalition with Bob as when she succeeds in forming a coalition with Carol – otherwise one of the potential agreements would require her to accept less than her outside option in that situation. Together with the Coase theorem, this fact ties down the three possible deals. In the case when the players are all risk neutral, we are led to the unsurprising conclusion that the coalition which forms will split the dollar fifty-fifty, leaving the odd-man-out with nothing.

The symmetry of the problem makes it impossible to say which of the three possible coalitions will form. However, the following noncooperative model breaks the symmetry by requiring that Alice, Bob, and Carol rotate in making payoff demands. When it is your turn to move, you may either accept any demand that has been made previously or else make a new demand of your own. The unique subgame-perfect equilibrium predicts that the very first opportunity to form a coalition will be seized by Alice and Bob. In order that their shares of the dollar approximate our cooperative prediction, the time interval between successive demands needs to be very small.

Core

What can be said about how coalitions form in more general situations? One proposal is that we should reject a payoff profile as a possible solution outcome if some coalition can object to it on the grounds that it is able to enforce an alternative payoff profile that all its members prefer. The set of all payoff profiles to which no such objection can be found is called the *core* of a cooperative game.

Economists like the idea because the core of a large enough market game approximates what will happen if buyers and sellers trade at whatever prices equate supply and demand. However, applying the idea to Odd-Man-Out in the case when all the players are risk neutral isn't very encouraging.

We have seen that one possible solution outcome in Odd-Man-Out is for Alice and Bob to form a coalition on the understanding that they will vote to split the dollar so that each gets 50 cents. But this outcome can't be in the core, because Bob and Carol can object that they are able to enforce an outcome that they both prefer by voting to split the dollar so that Bob gets 51 cents and Carol gets 49 cents. Since similar reasoning can be used to exclude any payoff profile whatever, the core of Odd-Man-Out is empty.

Condorcet paradox

The Marquis de Condorcet was an idealistic French revolutionary who discovered a similar problem when exploring possible voting systems. If Alice and Bob form a coalition that disadvantages Carol, she will offer whoever will listen a little more than they are currently getting. If Bob takes up Carol's offer and abandons Alice, then Alice will become the disadvantaged party, with an incentive to offer Carol a little more than she is currently getting. If Carol agrees, Bob will then approach Alice. And so on.

The results in real life can be devastating. For example, the border between England and Wales where I live was a battlefield for centuries. Powerful lords on the English side supposedly guarded the border or marches against raids by the Welsh tribes, but warfare was actually continuous as the Welsh, the King of England, and the local Marcher Lord shifted alliances to combine against whichever of the three was currently most powerful.

Condorcet's life didn't work out any better than the victims of the unstable social systems whose mechanics he succeeded in identifying. He had hoped to create a utopia by mathematical reasoning, but was sentenced to the guillotine instead.

Stable sets

Von Neumann and Morgenstern understood that Bob would be unwise to listen to Carol in Odd-Man-Out when she explains that he can get 51 cents by joining a coalition with her rather the 50 cents that Alice has promised him. If it is a good idea to dump Alice when he is approached by Carol, then it will be a good idea for Carol to dump him when she is approached by Alice.

To capture this idea, Von Neumann and Morgenstern invented a notion that is nowadays called a *stable set*. They argued that objections which aren't themselves possible solution outcomes should be ignored. Anything outside a stable set is still excluded because an objection from within the stable set can be found, but something inside a stable set need only be immune from objections within the stable set.

Their chief example was Odd-Man-Out when the players are all risk neutral. One stable set consists of the three possible outcomes in which the dollar is divided equally between two of the players. However, there are lots of other stable sets. For example, the set of all of outcomes in which Carol gets 25 cents and the rest of the dollar is split in all possible ways between Alice and Bob is stable.

It isn't easy to make sense of these new stable sets. Other game theorists disagree, but I think their appearance simply shows that the idea of a stable set isn't precise enough. So there are sometimes too many stable sets – but this is the least of our troubles. William Lucas found a cooperative game with many players that has no stable sets at all, and so there are also sometimes too few stable sets.

Shapley value

I was once summoned urgently to London to explain what the French government was talking about when it suggested that the costs of a proposed tunnel under the English Channel be allocated to countries in the European Union using the Shapley value. The latter is the brainchild of Lloyd Shapley, who was another of the brilliant group of graduate students who studied mathematics alongside John Nash at Princeton.

Shapley followed Nash's example by proposing a set of assumptions that define a unique prediction for the outcome of a cooperative game. However, unlike Nash, his assumptions apply not just to bargaining games with only two players, but to any cooperative game with 'transferable utility'. The leading case of interest is when the players are all risk neutral and the payoffs are measured in dollars. It can then be argued that everything that matters about a coalition is what I shall call the value of the coalition – the largest number of dollars that it can guarantee is available to be shared out among its members. These payoffs include any 'side payments' necessary to buy the loyalty of any member of the coalition who might think the grass looks greener elsewhere.

For example, in Odd-Man-Out, the value of each coalition with two players is one dollar. The value of the grand coalition of all three players is also one dollar. The value of a coalition with only one player is zero. The empty coalition with no players also has value zero.

The easiest way to find the Shapley value makes it explicit that it is intended as an *average* over all the possible ways that coalitions might form. Start with the empty coalition and add players until you get to the grand coalition. When Alice is added to a coalition, write down her marginal contribution to the coalition – the amount by which her inclusion increases the value of the coalition. The payoff assigned to Alice by the Shapley value is then the average of all her marginal contributions taken over all the possible ways in which the grand coalition can be assembled one player at a time.

Odd-Man-Out has three players, and so there are six ways of ordering the players: ABC, ACB, BAC, BCA, CAB, CBA. Alice's marginal contributions are respectively: 0, 0, 1, 0, 1, 0. So the Shapley value assigns Alice a payoff of $1/3$ of a dollar, which is what we argued she would get on average in the previous section on coalitions.

How useful is the Shapley value? I think there is no doubt of its relevance to cost-sharing exercises of the type proposed by the French government, but it doesn't fare too well when tested by the Nash program. Like much else in game theory, there remains a great deal about coalition formation that we do not yet understand.

Chapter 10
Puzzles and paradoxes

Feedback phenomena and human intuition are uncomfortable bedfellows. When people dislike where an equilibrium argument takes them, it is therefore unsurprising that they invent simpler arguments that lead to more palatable conclusions. However, the first principle of rational thought is never to allow your preferences to influence your beliefs.

Fallacies of the Prisoner's Dilemma

The fact that both players would be better off if they didn't play their equilibrium strategies in the Prisoner's Dilemma is said to be a paradox of rationality that requires resolution.

Categorical imperative

In colloquial language, Immanuel Kant's categorical imperative says that it is rational to do what you wish everybody would do. If this were true, it would be rational to cooperate in the Prisoner's Dilemma. But wishful thinking is never rational. It is a constant source of amazement to me that Kant is never held to account for proposing a rationality principle without giving any reasons why we should take it seriously.

Fallacy of the twins

Two rational people facing the same problem will necessarily choose the same action. So Alice and Bob will either both play *hawk* or both play *dove* in the Prisoner's Dilemma. Since Alice prefers the outcome (*dove, dove*) to (*hawk, hawk*), she should therefore choose *dove*.

The fallacy is attractive because it would be correct if Alice and Bob were genetically identical twins, and we were talking about what genetically determined behaviour best promotes biological fitness (see Kin selection in Chapter 8). But the relevant game wouldn't then be the Prisoner's Dilemma; it would be a game with only one player.

As is commonplace when looking at fallacies of the Prisoner's Dilemma, we are offered a correct analysis of the wrong game. The Prisoner's Dilemma is a two-player game in which Alice and Bob choose their strategies *independently*. The twins fallacy wrongly assumes that Bob will make the same choice as Alice whatever strategy she chooses. This can't be right, because Bob is supposedly rational and one of his two choices is irrational.

One can modify the assumptions of the fallacy so that Alice and Bob's strategies coincide only with some sufficiently high probability. The story told to justify such a correlation in their behaviour often kicks up enough dust to obscure the fact that any correlation at all implies that Alice and Bob aren't choosing independently. But if they don't choose independently, they aren't playing the Prisoner's Dilemma. Even if Alice and Bob's information were correlated, as hypothesized in Aumann's notion of a correlated equilibrium, they still wouldn't play *hawk*, because *hawk* is strongly dominated whatever the players may learn about other matters.

Myth of the wasted vote

A version of the twins fallacy is routinely trotted out at election time, when pundits argue that 'every vote counts' (see Mixed Nash equilibria, Chapter 2). If a wasted vote is one that doesn't affect the outcome of the election, then the only time that your vote can count is when only one vote separates the winner and the runner-up. If they are separated by two or more votes, then a change in your vote would make no difference at all to who is elected. However, an election for a seat in a national assembly is almost never settled by a margin of only one vote.

Here is a hypothetical example of an election even closer than the actual race between Bush and Gore in the United States in 2000. A reliable opinion poll says that the voters in a pivotal state who have made up their minds are split 51% to 49% in favour of Bush. The probability that a floating voter will go for Bush is just enough to ensure that he will beat Gore by 500 votes on average. Things look so close that Alice decides to vote. What are the chances that her vote will count – that the result would have been different if she had stayed home and watched the television?

With one million voters of whom 5% are undecided, Alice's vote would count only once in every 8,000 years, even if the same freakish circumstances were repeated every four years. But they won't be. The chances that the votes cast by floaters will almost balance those cast by the decided voters are infinitesimal. If the floaters in our example voted for Bush with the same frequency as the rest of the population, Alice's vote would count only once in every 20 billion billion years. No wonder no state has ever been decided by a single vote in a presidential election!

Naive folk imagine that to accept this argument is to precipitate the downfall of democracy. We are therefore told that you are wrong to count only the effect of your vote alone – you should instead count the total number of votes cast by all those people who think and feel as you think and feel, and hence will vote as you vote. If you have 10,000 such soul mates or *twins*, your vote wouldn't then be wasted, because the probability that an election will be decided by a margin of 10,000 votes or less is often very high. This argument is faulty for the same reason that the twins fallacy fails in the Prisoner's Dilemma. There may be large numbers of people who think and feel like you, but their decisions on whether to go out and vote won't change if you stay home and watch the television.

Critics sometimes accuse game theorists of a lack of public spirit in exposing this fallacy, but they are wrong to think that democracy would fall apart if people were encouraged to think about the realities of the election process. Cheering at a football game is a useful analogy. Few cheers would be raised if what people were trying to do by cheering was to increase the general noise level in the stadium. No single voice can make an appreciable difference to how much noise is being made when a crowd of people is cheering. But nobody cheers at a football game because they want to increase the general noise level. They shout words of wisdom and advice at their team even when they are at home in front of a television set.

The same goes for voting. You are kidding yourself if you vote because your vote has a significant chance of being pivotal. But it makes perfectly good sense to vote for the same reason that football fans yell advice at their teams. And, just as it is more satisfying to shout good advice rather than bad, so many game theorists think that you get most out of participating in an election by voting *as though* you were going to be the pivotal voter, even though you know the probability of one vote making a difference is too small to matter. A Kantian would assume that everyone is similarly strategic, but I prefer to use opinion polls when guessing the most likely way a tie might arise.

For example, Ralph Nader was the green candidate in the presidential election when Bush just beat Gore. I am hot on green issues, but I wouldn't have voted for Nader, because if there had been a tie, it would almost certainly have been between Bush and Gore. In Europe, such strategic voting will sometimes result in your voting for a minor party. The same pundits who tell you that every vote counts will also tell you that such a strategic vote is a wasted vote. But they can't be allowed to have it both ways!

Transparent disposition fallacy

This fallacy asks us to believe two doubtful propositions. The first is that rational people have the willpower to commit themselves in advance to playing games in a particular way. The second is that other people can read our body language well enough to know when we are telling the truth. If we truthfully claim that we have made an irrevocable commitment, we will therefore be believed.

If these propositions were correct, our world would certainly be very different! Charles Darwin's *Expression of the Emotions* would be wrong in denying that our involuntary facial muscles make it impossible to conceal our emotional state, and so actors would be out of a job. Politicians would be incorruptible. Poker would be impossible to play. Rationality would be a defence against drug addiction. However, the logic of game theory would still apply.

As an example, consider two possible mental dispositions called CLINT and JOHN. The former is a retaliating strategy named after the character played by Clint Eastwood in the spaghetti westerns (see Evolution of Cooperation, Chapter 8). The latter commemorates a hilarious movie I once saw in which John Wayne played the part of Genghis Khan. To choose the disposition JOHN is to advertise that you have committed yourself to play *hawk* in the Prisoner's Dilemma no matter what. To choose the disposition

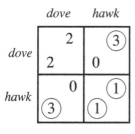

	dove	*hawk*
dove	2 / 2	(3) / 0
hawk	0 / (3)	(1) / (1)

Prisoner's Dilemma

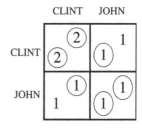

	CLINT	JOHN
CLINT	(2) / (2)	1 / (1)
JOHN	(1) / 1	(1) / (1)

Film Star Game

34. Transparent disposition fallacy

CLINT is to advertise that you are committed to play *dove* in the Prisoner's Dilemma if and only if your opponent is advertising the same commitment. Otherwise you play *hawk*.

If Alice and Bob are allowed to commit themselves transparently to one of these two dispositions, they won't be playing the Prisoner's Dilemma any more; they will be playing the Film Star Game of Figure 34 in which the players' strategies are CLINT and JOHN. If both players choose CLINT in the Film Star Game, they are then committed to playing *dove* in the Prisoner's Dilemma; otherwise they are committed to playing *hawk*.

As the circled payoffs show, CLINT is a (weakly) dominant strategy in the Film Star Game. So if Alice and Bob choose CLINT, they will be playing a Nash equilibrium that results in their cooperating in the Prisoner's Dilemma. Advocates of the transparent disposition fallacy think that this shows that cooperation is rational in the Prisoner's Dilemma. It would be nice if they were right that real-life games are really all Film Star Games of some kind – especially if one could choose to be Adam Smith or Charles Darwin rather than John Wayne or Clint Eastwood. But even then it wouldn't follow that rationality requires cooperating in the Prisoner's Dilemma. The argument shows only that it is rational to play CLINT in the Film Star Game.

Newcomb's paradox

Two boxes possibly have money inside. Alice is free to take either the first box or both boxes. If she cares only for money, what should she do? This seems an easy problem. If *dove* represents taking only the first box and *hawk* represents taking both boxes, then Alice should choose *hawk* because she then gets at least as much money as with *dove*.

However, there is a catch. It is certain that the second box contains one dollar. The first box contains either two dollars or nothing. The decision about whether there should be money in the first box is made by Bob, who knows Alice so well that he is always able to make a perfect prediction of what she will do. Like Alice, he has two choices, *dove* and *hawk*. His dovelike choice is to put two dollars in the first box. His hawkish choice is to put nothing in the first box. His motivation is to catch Alice out. He therefore plays *dove* if he predicts that Alice will choose *dove*. He plays *hawk* if he predicts that Alice will choose *hawk*.

Choosing *hawk* doesn't look so good for Alice now. If she chooses *hawk*, Bob predicts her choice and puts nothing in the first box, so that Alice gets only the single dollar in the second box. But if Alice chooses *dove*, Bob will predict her choice and put two dollars in the first box for her to pick up.

The Harvard philosopher Robert Nozick created a craze in his profession (aptly described as Newcombmania) by claiming that Newcomb's paradox shows you can sometimes maximize your payoff by playing a strongly dominated strategy. He could equally well have argued that it shows $2 + 2 = 5$, since anything can be deduced from a contradiction. The contradiction in Newcomb's paradox consists in assuming the existence of a game in which:

1. Alice moves after Bob.
2. Bob knows Alice's choice.
3. Alice has more than one choice.

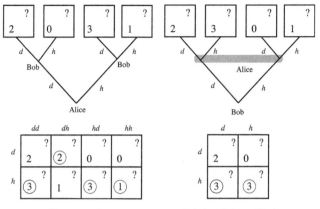

	dd	dh	hd	hh
d	2 ?	② ?	0 ?	0 ?
h	③ ?	1 ?	③ ?	① ?

	d	h
d	2 ?	0 ?
h	③ ?	③ ?

Alice moves first   Bob doesn't know Alice's move

35. Two attempts to satisfy Newcomb's requirements. The information set in the right-hand game indicates that Alice doesn't know Bob's prediction. The payoff tables underneath each game tree are the relevant strategic forms

Figure 35 shows two attempts to create such a game without being specific about Bob's payoffs; the game on the left fails item 1 on the list, and that on the right fails item 2. We can satisfy both item 1 and item 2 by offering Alice only one choice in the right-hand game, but then we fall foul of item 3.

When arguing that Alice must play *dove* to maximize her payoff, Nozick assumes that Bob will play *dh* in the left-hand game. That is to say, Bob will predict *d* when Alice plays *d* and *h* when she plays *h*. However, Alice's strategy *d* isn't dominated in the left-hand game. To argue that Alice's strategy *d* is dominated, one has to appeal to the right-hand game. But it isn't paradoxical that Alice might play differently in different games.

One can muddy the waters by giving up the requirement that Bob can predict Alice's behaviour *perfectly*. We can then create a game in which the three requirements of Newcomb's paradox are satisfied by introducing chance moves into the right-hand game that remove Alice's opportunity to choose differently from Bob some of the time. But no amount of juggling with the parameters will make it optimal to play a strongly dominated strategy!

Surprise test paradox

The British telecom auction that raised $35 billion has been mentioned several times. Everybody was surprised at this enormous amount – except for the media experts, who finally got the figure roughly right by predicting a bigger number whenever the bidding in the auction falsified their previous prediction. Everybody can see the fraud perpetrated by the media experts on the public in this story, but the fraud isn't so easily detected when it appears in one of the many versions of the surprise test paradox, through which most people first learn of backward induction.

Alice is a teacher who tells her class that they are to be given a test one day next week, but the day on which the test is given will come as a surprise. Bob is a pupil who works backward through the days of the coming school week. If Alice hasn't set the test by the time school is over on Thursday, Bob figures that she will then have no choice but to set the test on Friday – this being the last day of the school week. So if the test were given on Friday, Bob wouldn't be surprised. Bob therefore deduces that Alice can't plan to give the test on Friday. But this means that the test must be given on Monday, Tuesday, Wednesday, or Thursday. Having reached this conclusion, Bob now applies the backward induction argument again to eliminate Thursday as a possible day for the test. Once Thursday has been eliminated, he is then in a position to eliminate Wednesday. Once he has eliminated all the days of the school week by this method, he sighs with relief and makes no attempt to study over the weekend. But then Alice takes him by surprise by setting the test first thing on Monday morning!

This isn't really a paradox at all, because Bob shouldn't have been so quick to sigh with relief. If the backward induction argument is correct, then Alice's two statements are inconsistent, and so at least one of them must be wrong. But why should Bob assume that the wrong statement is that a test will be given, and not that the test will come as a surprise? This observation is usually brushed aside, because what people really want to hear about is whether the backward induction argument is right. But what they should be asking is whether backward induction has been applied to the right game.

In the game that people imagine is being analysed, Eve chooses one of five days on which to hold the test, and Bob predicts which of the five days she will choose. If his prediction is wrong, then he will be taken by surprise. The solution of this version of Matching Pennies is that Alice and Bob both choose each day with equal probability. Bob is then surprised four times out of five.

This isn't the conclusion we reached before, because the surprise test paradox applies backward induction to a game in which Bob is always allowed to predict that the test will be today, even though he may have wrongly predicted that it was going to take place yesterday. In this bizarre game, Bob's optimal strategy is therefore to predict Monday on Monday, Tuesday on Tuesday, Wednesday on Wednesday, Thursday on Thursday, and Friday on Friday. No wonder Bob is never surprised by having the test occur on a day he didn't predict!

The surprise test paradox has circulated ever since I can remember. Occasionally it gets a new airing in newspapers and magazines. It has even been the object of learned articles in philosophical journals. The confusion persists because people fail to ask the right questions. One of the major virtues of adopting a systematic formalism in game theory is that asking the correct questions becomes automatic. You then don't need to be a genius like Von Neumann to stay on the right track. His formalism does the thinking for you.

Common knowledge

Why do we attach so much importance to eye contact? I think the reason is that something becomes common knowledge only if it is implied by an event that couldn't have occurred without everybody knowing it. For example, if Alice and Bob observe each other observing that Carol has a dirty face, then it becomes common knowledge between Alice and Bob that Carol has a dirty face. Similarly, when two people look each other in the eye, it becomes common knowledge between them that they are aware of each other as individuals.

Three old ladies

Alice, Beatrice, and Carol are three respectable ladies at a midwestern county fair. Each has a dirty face, but nobody is blushing, although a respectable lady who was conscious of appearing in public with a dirty face would surely do so. It follows

36. Three midwestern ladies

that none of the ladies knows that her own face is dirty, although each can clearly see the dirty faces of the others.

Midwestern clergymen always tell the truth, and so the ladies pay close attention when a local minister announces that one of the ladies has a dirty face. After his announcement, one of the ladies blushes. How come? Didn't the minister simply tell the ladies something they knew already?

To understand what the minister added to what the ladies already knew, we need to look at the chain of reasoning that leads to the conclusion that at least one of the ladies must blush. If neither Beatrice nor Carol blushes, Alice would reason as follows:

> *Alice:* Suppose that my face were clean. Then Beatrice would reason as follows:
>
>> *Beatrice:* I see that Alice's face is clean. Suppose that my face were also clean. Then Carol would reason as follows:
>>
>>> *Carol:* I see that Alice and Beatrice's faces are clean. If my face were clean, nobody's face would be dirty. But the minister's announcement proves otherwise. So my face is dirty, and I must blush.
>>
>> *Beatrice:* Since Carol hasn't blushed, my face is dirty. So I must blush.
>
> *Alice:* Since Beatrice hasn't blushed, my face is dirty. So I must blush.

So what did the minister add to what the ladies already knew? For Alice's reasoning to work, she needed to know that Beatrice knows that Carol knows that Alice and Beatrice know that someone has a dirty face. All these knowings became possible only after the minister's announcement makes it common knowledge that someone has a dirty face. It is then not only true that Alice, Beatrice, and Carol know that one of them has a dirty face;

they all know that they all know that they all know that they know it.

A coordination paradox

Is a magnificent beard necessary to make advances in interactive epistemology? The only evidence I have to offer is that the bearded Princeton philosopher David Lewis shares the credit for recognizing the importance of common knowledge in game theory with the equally hirsute Bob Aumann. But what are we to make of Lewis's claim that a convention can't be operational unless it is common knowledge that the players are planning to use it?

For something to become common knowledge, we need an equivalent of the tactless clergyman in the story of the three midwestern ladies. But no such clergyman is usually to be found. Nearly all the conventions we use in daily life therefore fail Lewis's test. So how come they seem to work so well?

Computer scientists worried about the implications for distributed systems illustrate the problem by telling a story about two Byzantine generals trying to coordinate an attack on an enemy army that lies in a valley between them, but I prefer a less dramatic example.

Alice and Bob want to get together tomorrow in New York. Alice emails the suggestion that they meet at Grand Central Station at noon. Bob emails a confirmation. This exchange would be adequate for most of us, but Lewis would object that the agreement isn't common knowledge because Bob doesn't know that Alice received his confirmation. She should therefore email to confirm that she received his confirmation. Bob should then email a confirmation of her confirmation, and so on. Since there is always a small probability that an email message won't be received, their attempt to agree on a convention will never become common knowledge.

But why should a convention have to be common knowledge to be operational? Ariel Rubinstein studied this question by analysing a new Email Game in which Alice and Bob's Meeting Game is replaced by the Stag Hunt Game of Chapter 4. The default convention is for Alice and Bob to play *dove* in the Stag Hunt Game, but every so often the labels of both their strategies get reversed, so that choosing *dove* will result in *hawk* actually being played. Only Alice observes when this happens. She sends an email message to Bob saying that they should play *hawk* on this occasion rather than *dove*. He automatically sends a confirmation. She automatically sends a confirmation of his confirmation, and so on.

A strategy in the Email Game says whether *dove* or *hawk* should be played depending on the number of messages a player has received. We can then short-circuit the common knowledge question by asking whether there is a Nash equilibrium of the Email Game in which Alice and Bob always succeed in coordinating on the equilibrium they both prefer in the Stag Hunt Game. Rubinstein's answer seems to confirm Lewis's intuition. The only Nash equilibrium in the Email Game in which Alice and Bob play *dove* when no message is sent requires that they *always* play *dove* no matter how many messages they may receive.

However, the picture changes when we allow Alice and Bob to choose whether or not to send or receive messages. The modified Email Game then has many Nash equilibria, the most pleasant of which requires that both players play *hawk* whenever Alice proposes doing so and Bob says OK – as when friends agree to meet in a coffee shop. But there are other Nash equilibria in which the players settle on *hawk* only after a long exchange of confirmations of confirmations. Hosts of polite dinner parties suffer from such equilibria when their guests start moving with glacial slowness towards the door at the end of the evening, stopping every inch or so in order that the host and the guest can repeatedly assure each other that departing at this time is socially acceptable to both sides.

The common-sense conclusion is that conventions don't need to be common knowledge to work. Most conventions are established by the forces of cultural evolution. Sometimes evolutionary stability considerations make it possible to eliminate some Nash equilibria. In the modified Email Game, one might hope that such considerations would eventually eliminate the equilibria that generate 'long goodbyes' after dinner parties, but the prognosis isn't good. Ironically, only Rubinstein's equilibrium, in which Alice and Bob play *dove* no matter what happens, fails to pass an appropriate evolutionary stability test.

Monty Hall problem

Alice is a contestant in an old quiz show run by Monty Hall. She must choose from three boxes, only one of which contains a prize. Monty knows which box contains the prize, but Alice doesn't. After she chooses *Box 2*, Monty opens one of the other boxes that he knows to be empty. Alice then has the opportunity to change her mind about her choice of box. What should she do?

People usually say it doesn't matter. They reason that Alice's probability of winning when she chose *Box 2* was 1/3 because there was then an equal chance of the prize being in any of the three boxes. After another box is shown to be empty, the probability that *Box 2* contains the prize goes up to 1/2, because there is now an equal chance that the prize is in one of the two unopened boxes. If Alice switches boxes, her probability of winning will therefore still be 1/2. So why bother changing?

Marilyn Vos Savant apparently has the highest IQ ever recorded. When she explained in *Parade* magazine that Alice should always switch boxes, various self-appointed mathematical gurus laughed her to scorn, but she was right.

The probability that the prize is either in *Box 1* or *Box 3* is 2/3. If she switches to whichever of these boxes isn't opened, Alice will therefore win with probability 2/3.

This argument is deceptively easy. Even top mathematicians sometimes fail to see why Monty's action conveys so much information to Alice. After all, it wouldn't have conveyed any useful information at all if he had opened a box at random that just happened to be empty – but he deliberately chose a box that he knew to be empty.

However, you don't need to have the highest IQ ever recorded to get the answer right if you are willing to let Von Neumann do your thinking for you. Figure 37 shows the game that Alice and Monty

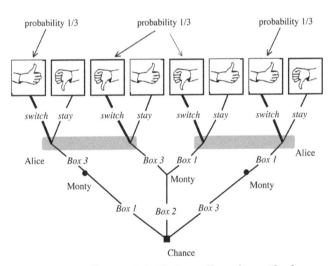

37. The Monty Hall Game. Only Alice's payoffs are shown. The chance move is shown as a square. Alice's information sets show that she doesn't know which box contains the prize, but she does know which box Monty opens. Her switching choice is thickened. The figure shows that whatever strategy Monty chooses, Alice wins with probability 2/3 if she switches

are playing. It doesn't matter what Monty's payoffs are, but we might as well assume that he wants Alice to lose. A chance move first puts the prize in one of the boxes. Monty then decides whether to open *Box 1* or *Box 3*. (He only has a genuine choice when the prize is actually in *Box 2*.) Alice then chooses whether to stay with *Box 2* or to switch to whichever of *Box 1* or *Box 3* Monty didn't open.

There is now no need to think at all. If Alice always switches, the figure makes it impossible not to recognize that she wins when the prize is in *Box 1* or *Box 3* and loses when the prize is in *Box 2*. So she wins with probability 2/3.

References and further reading

Chapter 1

Ken Binmore, *Playing for Real* (New York: Oxford University Press, 2007). This textbook on game theory is light on mathematics.

Ken Binmore, *Natural Justice* (New York: Oxford University Press, 2005). Why game theory matters in ethics.

Colin Camerer, *Behavioral Game Theory* (Princeton: Princeton University Press, 2003). Some aspects of game theory work well in the laboratory, and some don't. This book surveys the evidence, and looks at possible psychological explanations of deviations from the theory.

John Rawls, *A Theory of Justice* (Oxford: Oxford University Press, 1972). Rawls is often said to be the greatest moral philosopher of the 20th century. He refers to the maximin principle as the difference principle when proposing it as a rational substitute for maximizing average utility.

John Maynard Smith, *Evolution and the Theory of Games* (Cambridge: Cambridge University Press, 1982). This beautiful book introduced game theory to biology.

Barry Nalebuff and Avinash Dixit, *Thinking Strategically* (New York: Norton, 1991). A book-club choice, it contains many examples of game theory in action, both in business and in everyday life.

Sylvia Nasar, *A Beautiful Mind* (New York: Simon and Schuster, 1998). A best-selling biography of John Nash.

Alvin Roth and John Kagel, *Handbook of Experimental Game Theory* (Princeton: Princeton University Press, 1995). The survey by John Ledyard documents the immense amount of data supporting the claim that experienced subjects seldom cooperate in the Prisoner's Dilemma.

John Von Neumann and Oskar Morgenstern, *The Theory of Games and Economic Behavior* (Princeton: Princeton University Press, 1944). Not a best-seller. Their theory of utility appears in an appendix.

Chapter 2

Robert Aumann, *Lectures on Game Theory* (Boulder, CO: Westview Press Underground Classics in Economics, 1989). The classroom notes of one of the great game theorists.

Ken Binmore, *Does Game Theory Work?* (Cambridge, MA: MIT Press, 2007). This book includes my own experiment on zero-sum games and references to others.

Steve Heine, *John von Neumann and Norbert Wiener* (Cambridge, MA: MIT Press, 1982). I write 'Von Neumann' rather than 'von Neumann' because one gets into trouble in some parts of the German-speaking world for according him the title that his father bought from the Hungarian government.

J. D. Williams, *The Compleat Strategyst* (New York: Dover, 1954). A delightful collection of simple two-person, zero-sum games.

Chapter 3

Robert Aumann, 'Interactive Epistemology', *International Journal of Game Theory*, 28 (1999): 263–314.

Martin Gardner, *Mathematical Diversions* (Chicago: University of Chicago Press, 1966) and *Hexaflexagons* (Chicago: University of Chicago Press, 1988). These books gather together many delightful games and brainteasers from the author's long-standing column in *Scientific American*.

Robert Gibbons, *Game Theory for Applied Economists* (Princeton: Princeton University Press, 1992). An unfussy introduction to game theory, with an orthodox treatment of refinements.

David Lewis, *Counterfactuals* (Cambridge, MA: Harvard University Press, 1973).

Larry Samuelson, *Evolutionary Games and Equilibrium Selection* (Cambridge, MA: MIT Press, 1997). This includes our paper on the replicator dynamics in the Ultimatum Game.

Chapter 4

Steven Brams, *Superior Beings: If They Exist, How Would We Know? Game Theoretic Implications in Omniscience, Omnipotence, Immortality and Comprehensibility* (New York: Springer Verlag, 1983).

John Harsanyi and Reinhard Selten, *A General Theory of Equilibrium Selection in Games* (Cambridge, MA: MIT Press, 1988).

David Hume, *A Treatise of Human Nature* (Oxford: Clarendon Press, 1978; first published 1739). Arguably the greatest work of philosophy ever.

David Lewis, *Conventions* (Princeton: Princeton University Press, 1969).

Thomas Schelling, *The Strategy of Conflict* (Cambridge, MA: Harvard University Press, 1960). Schelling once bravely told a large audience of game theorists that game theory had contributed nothing whatever to the theory of focal points – except perhaps the idea of a payoff table!

Thomas Schelling, *Micromotives and Macrobehavior* (New York: Norton, 1978). Schelling's Solitaire and a lot more.

Brian Skyrms, *The Stag Hunt and the Evolution of the Social Structure* (Cambridge: Cambridge University Press, 2003).

Peyton Young, *Individual Strategy and Social Structure: An Evolutionary Theory of Institutions* (Princeton: Princeton University Press, 1998).

Chapter 5

Bob Axelrod, *Evolution of Cooperation* (New York: Basic Books, 1984). This book sold the world on the idea that reciprocity matters.

'Review of *The Complexity of Cooperation* by Ken Binmore', *Journal of Artificial Societies*, http://jasss.soc.surrey.ac.uk/1/1/review1.html. The book is a sequel to Axelrod's *Evolution of Cooperation*; the review assesses his reiterated claims for TIT-FOR-TAT. See also Karl Sigmund's *Games of Life* (Chapter 8 below).

Joe Heinrich *et al.* (eds), *Foundations of Human Sociality: Economic Experiments and Ethnographic Evidence from Fifteen Small-Scale Societies* (New York: Oxford University Press, 2004). An attempt to refute the repeated-game explanation of social norms that backfired. The paper by the anthropologist Jean Ensminger is particularly instructive.

George Mailath and Larry Samuelson, *Repeated Games and Reputations: Long-Term Relationships* (New York: Oxford University Press, 2006). Folk theorems with imperfect monitoring for mathematicians.

Bob Trivers, *Social Evolution* (Menlo Park, CA: Cummings, 1985). Reciprocity and much else in animal societies.

Chapter 6

Helena Cronin, *The Ant and the Peacock* (Cambridge: Cambridge University Press, 1991).

John Harsanyi, *Rational Behaviour and Bargaining Equilibrium in Games and Social Situations* (Cambridge: Cambridge University Press, 1977).

Roger Myerson, *Game Theory: Analysis of Conflict* (Cambridge, MA: Harvard University Press, 1991).

Chapter 7

Ken Binmore and Paul Klemperer, 'The Biggest Auction Ever: The Sale of British 3G Licences', *Economic Journal*, 112 (2002): C74–C96.

R. Cassady, *Auctions and Auctioneering* (Berkeley, CA: University of California Press, 1967). Lots of good stories.

Paul Klemperer, *Auctions: Theory and Practice* (Princeton: Princeton University Press, 2004).

Paul Milgrom, *Putting Auction Theory to Work* (Cambridge: Cambridge University Press, 2004).

Chapter 8

John Alcock, *The Triumph of Sociobiology* (Oxford: Oxford University Press, 2001). Sociobiologists aren't the intellectual fascists they have been painted. Aside from offering wonderful examples of real sociobiology in action, this book lays bare the dishonest campaign of vilification directed at Edward Wilson and his followers by Gould, Lewontin, and other politically motivated polemicists.

Ken Binmore and Larry Samuelson, 'Evolutionary Stability in Repeated Games Played by Finite Automata', *Journal of Economic Theory*, 57 (1992): 278–305.

Richard Dawkins, *The Selfish Gene* (Oxford: Oxford University Press, 1976). One of the great works of popular science.

Peter Hammerstein, *Genetic and Cultural Evolution of Cooperation* (Cambridge, MA: MIT Press, 2003).

William Hamilton, *The Narrow Roads of Geneland* (Oxford: Oxford University Press, 1995). A collection of some of Bill Hamilton's path-breaking papers in evolutionary biology. The papers themselves are not easy reading for the general reader, but the linking remarks are a fascinating social commentary on how it was to be a graduate student in the old days, doing work so original that the academic establishment was unable to appreciate its value.

John Maynard Smith, *Evolution and the Theory of Games* (Cambridge: Cambridge University Press, 1984). Many wonderful examples.

Karl Sigmund, *Games of Life: Explorations in Ecology, Evolution and Behaviour* (Harmondsworth: Penguin Books, 1993). Among other delights, this book reports on some of the author's computer simulations with Martin Nowack. Their name for TIT-FOR-TAT is PAVLOV (see Chapter 5).

James Watson, *The Double Helix: A Personal Account of the Discovery of the Structure of DNA* (New York: Touchstone, 1968).

Vero Wynne-Edwards, *Animal Dispersion in Relation to Social Behaviour* (Edinburgh: Oliver and Boyd, 1962).

Chapter 9

Ken Binmore, *Playing for Real* (New York: Oxford University Press, 2007). Four chapters are devoted to bargaining issues.

Ken Binmore, *Natural Justice* (New York: Oxford University Press, 2005). This book explains why I side with Rawls rather than Harsanyi on the implications of using the original position to make fairness judgements.

Roger Fisher *et al.*, *Getting to Yes* (London: Houghton Mifflin, 1992). This best-seller argues that good bargaining consists of insisting on a fair deal. Thinking strategically is dismissed as a dirty trick!

Howard Raiffa, *The Art and Science of Negotiation* (Cambridge, MA: Harvard University Press, 1982).

Chapter 10

Ken Binmore, *Playing Fair: Game Theory and the Social Contract I* (Cambridge, MA: MIT Press, 1995). Chapter 3 discusses more fallacies of the Prisoner's Dilemma that circulate in the philosophical literature.

Bob Frank, *Passions with Reason* (New York: Norton, 1988). An economist makes a case for the transparent disposition fallacy.

David Lewis, *Conventions: A Philosophical Study* (Cambridge, MA: Harvard University Press, 1969).

J. E. Littlewood, *Mathematical Miscellany* (Cambridge: Cambridge University Press, 1953). I was a schoolboy when I first came across the paradox of three old ladies in this popular work by one of the great mathematicians.

Game Theory

Index

N

O

P

R

LOGIC
A Very Short Introduction
Graham Priest

Logic is often perceived as an esoteric subject, having little to do with the rest of philosophy, and even less to do with real life. In this lively and accessible introduction, Graham Priest shows how wrong this conception is. He explores the philosophical roots of the subject, explaining how modern formal logic deals with issues ranging from the existence of God and the reality of time to paradoxes of self-reference, change, and probability. Along the way, the book explains the basic ideas of formal logic in simple, non-technical terms, as well as the philosophical pressures to which these have responded. This is a book for anyone who has ever been puzzled by a piece of reasoning.

'a delightful and engaging introduction to the basic concepts of logic. Whilst not shirking the problems, Priest always manages to keep his discussion accessible and instructive.'

Adrian Moore, St Hugh's College, Oxford

'an excellent way to whet the appetite for logic. . . . Even if you read no other book on modern logic but this one, you will come away with a deeper and broader grasp of the *raison d'être* for logic.'

Chris Mortensen, University of Adelaide

www.oup.co.uk/isbn/0-19-289320-3

DARWIN
A Very Short Introduction
Jonathan Howard

Darwin's theory of evolution, which implied that our ancestors were apes, caused a furore in the scientific world and beyond when *The Origin of Species* was published in 1859. Arguments still rage about the implications of his evolutionary theory, and scepticism about the value of Darwin's contribution to knowledge is widespread. In this analysis of Darwin's major insights and arguments, Jonathan Howard reasserts the importance of Darwin's work for the development of modern science and culture.

> 'Jonathan Howard has produced an intellectual *tour de force*, a classic in the genre of popular scientific exposition which will still be read in fifty years' time'
>
> **Times Literary Supplement**

www.oup.co.uk/isbn/0-19-285454-2